D0960711

I'll Take Your Questions Now

I'll Take Your Questions Now

WHAT I SAW AT THE TRUMP WHITE HOUSE

Stephanie Grisham

HARPER

An Imprint of HarperCollins*Publishers*

HarperCollins books may be purchased for educational, business, or sales promotional use. For information, please email the Special Markets Department at SPsales@harpercollins.com.

FIRST EDITION

Library of Congress Cataloging-in-Publication Data has been applied for.

ISBN 978-0-06-314293-0

21 22 23 24 25   LSC   10 9 8 7 6 5 4 3 2 1

Kurtis and Jake . . . I love you both more than anything else and am SO PROUD of the people you are. Thank you for making me a mom, it is the best job I've ever had and has given me happiness I didn't anticipate or maybe deserve. We have so many memories left to make.

Dana . . . I hope you are proud.

Gramps and my pops . . . The greatest and most loved men I have ever known.

Heather, Troy, and Erik . . . Thank you for defending me from snakes and (re)teaching me about real life.

My family near and far . . . I love all of you and am grateful for whatever part you played in my life.

My circle of trust . . . You know who you are, and I am beyond blessed to count you as friends. Thank you for your support, especially in the last year.

Don't ever regret knowing someone in your life. Good people will give you happiness, bad people will give you experience, while the worst people will give you a lesson and the best people will always give you memories.

—Author Unknown

Contents

Author's Note

This book is based on my own memories, perceptions, and views of the last six years, in addition to other supporting information. I was present at most of the events described, and in the few instances where I was not, I relied on the contemporaneous statements of colleagues. Dialogue is reconstructed from my memory, what I noted at the time, or what others, such as the president and first lady, told me.

I'll Take Your Questions Now

Introduction

You can't go back and change the beginning, but you can start where you are and change the ending.

—ATTRIBUTED TO C.S. LEWIS

I could start this book any number of ways. I could tell you about the way Vladimir Putin toyed with the president of the United States by using a very attractive translator at one of their meetings or when there was a fight with Chinese security during a trip to Beijing and the nuclear football was in the middle of it. Or when President Trump, a solemn, intense look crossing his face, leaned toward me on board Marine One to ask me the most pressing question on his mind: "Are your teeth real?" Or when Mrs. Trump enlisted her staff to reenact a White House event at the new tennis court she'd had remodeled so she could get a better photo of herself for her photo albums. Those albums—those many, many albums—you are going to be hearing about them in this book. Or election night 2020, when President Trump was being advised by a bunch of aides and randos—*Hey there, Laura Ingraham!*—on how to respond to what everyone knew, but pretended they didn't, was a defeat. Spoiler alert: that didn't work out too well. Then there were all of Jared Kushner's schemes and seemingly shady dealings and "the Princess" (as the first lady and I, among others, called Ivanka Trump) trying to grab all the face time with Queen Elizabeth, her "fellow royal."

For nearly six years I was as close to the Trump family as anyone could be. I saw everything. I saw it all. I saw so much that I forgot some of it until I started writing this book—which you will see is part chronicle of a chaotic administration, part therapy session, and part personal reckoning. I have to get this all out so I can process, in my own mind, what the hell happened.

But maybe the best way to start this story is with the ending.

On January 6, 2021, after several botched attempts, I finally left the Trump administration. To put it another way, I resigned as not only the first lady's chief of staff but one of the longest-serving "hangers-on" with the Trump family over the past six years. They didn't like that too much. Immediately my official phone was cut off and I was placed on a "Do Not Admit" list at the White House. That was something. That crowd couldn't do anything fast or efficiently—except when it came to petty revenge. We were really very good at that.

I WILL BE CALLED many names for writing this book. Traitor. Low-level staffer. Weak. Dishonest. Ineffective. A complete failure. Anonymous sources will label me an alcoholic or someone with addiction issues (I know because this has already happened with the press). Statements will go out that I want attention and am self-serving, a person who wants to profit off the Trump family. More anonymous sources will follow up to say that I was never loyal and leaked to the press—the biggest sin of all in the Trump administration. The president and maybe even the first lady will pretend they hardly knew me or they will sue me—or both. I know all of this, because that is what we did to everyone else who decided to tell their truth or who stood up against things they thought were wrong. It is what I did, personally, to many people because it is the Trump formula: when you're in, you're in, and when you're out—you're denied and then destroyed. It's something of poetic justice, I guess, that I was once a destroyer myself.

I had been with the Trump family almost from the beginning of their political crusade—maybe the only person in history who worked at the most senior levels at the same time for the president and his wife. I spent nearly every holiday with them and flew with them all over the world. I was with the boss in private meetings with foreign leaders where he would say the most bonkers things and also, to be fair, often fight hard for our country. I was his spokeswoman during the (first) impeachment when he tried to get me to humiliate myself in front of the press and I had to trick him out of it. I was involved in standoffs with members of the Trump family over one thing or another. And I became a confidante of his wife, along with her two children: her son and her photo collection. (She loved her son the most.)

BUT BACK TO THE horror that was January 6, 2021.

Each time I tried to leave the White House for one reason or another, the first lady, whom I felt close to and protective of, talked me out of it. She was very persuasive, including with her husband. At least sometimes. But she couldn't talk me out of leaving that day.

I'd had a sick feeling in my stomach all morning, which was unusual. After five years of being immersed in all things Trump, you grew immune to chaos. Even in a weird way comfortable with it. There became something almost soothing about the knowledge that you are simply captive to a whirlwind you can't ever control—so you sit back, focus on staying alive, and let the winds take you where they will. In fact, the longer I was there, a sick pride built in me that I had outlasted so many others. What should have been warning signs were twisted into badges of honor.

Still, there was something different in the air this time. I just felt it.

Then I saw it.

I watched as a group of very angry and frenzied people scaled the walls of our Capitol, egged on by the president and his months of conspiracy mongering (thanks, Rudy). They were on their way to behead Mike Pence, kill AOC, or arrest Nancy Pelosi. Who the hell knew? It was scary and sad and sick and in some ways inevitable, I guess. *Of course* this was how it was all going to end.

Still, I had a job to do. I was the first lady's top aide, so I texted her. She had often been the sensible one. She had been the first to react to the violent clashes in Charlottesville, and how many times had we worked together and occasionally even succeeded in getting the president to dial back his rhetoric, to calm one situation or another? I'd lost count. I suggested that we send out a tweet immediately calling for an end to the violence and urging people to stand down. At 1:25 p.m., I texted her, "Do you want to tweet that peaceful protests are the right of every American, but there is no place for lawlessness and violence?"

From the Executive Residence of the White House, literally one minute later and while she was preparing a photo shoot of a new rug she had selected—yes, you read that right—Melania Trump sent me back a one-word response: "No."

I looked at that for a minute. A long minute. Then I looked at the TV again.

"No."

So many times over the years I had pushed back on the caricature of Melania Trump as some sort of Marie Antoinette, cold and cruel and oblivious to the little people over whom her husband ruled. That was not the woman I knew and admired and even had affection for. But that day, as the city of Washington, DC, descended into violence that had once seemed unimaginable, I finally saw the doomed French queen. Dismissive. Defeated. Detached.

It broke me. I took a breath and waited another minute. You learned to do that in the Trump White House: make sure you are grounded and not acting out of the moment. Then I resigned. I sent her an email and cc'd her senior adviser so I couldn't take it back or be talked out of it. I was done.

ONE THING I'M PROUD of that day was that I was the first to resign over the siege on the Capitol. And perhaps news of my departure emboldened others, or so I'd like to think. Shortly after news of my resignation leaked out, others resigned. Then Secretary of Transportation Elaine Chao quit—a big deal, since she was Senate Majority Leader Mitch McConnell's wife. Then Secretary of Education Betsy DeVos. Then another. Then another.

Of course the question had to be asked—and it was by many: Why did we wait so long to leave? Why did *I* wait so long? I had stayed through *Access Hollywood*, impeachment, family separation, Charlottesville, accusations of rape and misconduct, and a million other things that had whirled by me in a blur.

I'm not going to have an answer that will satisfy many people, and I know it's useless to try. I was a White House press secretary in the Trump administration. I know how to take my lumps. I know the spin doesn't work. I'll own the things I took part in and why I did them. Some people won't ever understand. Others will pretend that they'd never make the same choices in my circumstances. Nothing I can do about that. But I'd like to have an answer at least for myself and for those who care to hear it. It's also my chance to talk about my two biggest regrets, which may surprise you in their simplicity. People who want to journey with me through this book and learn about what I saw in the Trump White House should at least have a sense of where I was coming from.

The first thing I'd say about why I stayed is that working in the White House had been my lifelong dream. Being White House press secretary, for any president, was my greatest career ambition. I thought of how proud my grandpa would be. He had worked in the Ronald Reagan administration and was one of the few men in my life who looked out for me and who I loved and admired. I had a chance to go and serve my country—probably the only chance I'd ever get—and I took it. No regrets there. Grandpa would have my back.

And let's face it, *somebody* had to work in the Donald Trump White House. The president's opponents don't like this line of argument, but the administration was staffed by many dedicated people who, out of a sense of duty, were doing their best to make the government function under an inexperienced and unpredictable executive. Remember that even the resistance hero Mitt Romney once sought a job with Trump, auditioning for secretary of state even though he had denounced Trump during the campaign. I watched Romney "interview" at Trump's golf club in Bedminster, New Jersey; then I was there for the memorable dinner in New York City, when the Trump people dangled the job in front of him just to see if he would dance. (He danced!)

I hadn't initially supported Trump; I'd worked for Romney in 2012 and a number of candidates in early 2015 before I settled on Trump. But once I was in it, I was in it. People don't want to admit it now, but there was something refreshing about Donald Trump when he first arrived on the political scene, especially compared to all the other politicians, who were saying one predictable, poll-tested, lame thing after another. He was bold and poked at convention. He challenged dumb rules that people had just lived with for no reason. He said things people thought but never said. He took positions that no Republican had ever taken, including some shared by Bernie Sanders. During the primary race, he attacked George W. Bush and the Iraq War and railed against the deployment of our military into "endless wars" overseas. He talked about deserted industrial towns in middle America that both parties had essentially abandoned. He made weird asides and pop culture rants that no political candidate would ever utter. Once, on Air Force One, I was sitting with him in his cabin, and for whatever reason, maybe he had just read something or seen his face on TV, but Canadian prime minister Justin Trudeau popped into the president's head. Trump looked at me. "Are you okay if I say this?" That was always a troubling question. Who knew what was going to come out of his mouth? Sure, I nodded. "Trudeau's mom. She fucked all of the Rolling Stones." (Don't ask me where he came up with that one.) Yes, he could be offensive and over the top, but that overlooks part of his appeal. He was also deliberately outrageous, and he was funny. He'd had a hit TV show for years for a reason.

But as a candidate for president, Donald Trump wasn't so entertaining anymore—at least not to his former fans in the media, who had written puff pieces about him for years and soaked up his exaggerations and larger-than-life persona. Suddenly he could do nothing right. Every day, it seemed, brought a new revelation, a new scandal, a new accusation. Not all of them had equal weight. Not all of them were proven. Not all of them were true. The endless attacks on Trump actually had the opposite effect from what the more partisan critics may have hoped for: it became impossible to keep up with it all. We became hardened to it. It all started to fall on deaf ears. When you are buffeted by daily controversies and grievances and crises, and sometimes just false information, those who are fighting back tend to form a tight bond. I felt, in a way, part of a family of misfits who clung together, fighting against the established order. We had an imperfect leader at the top, and none of us wanted to see how imperfect he actually was. Just as his critics never wanted to give Trump credit for anything, we didn't want to give any credit to the critics who hated us. Even when they were right.

So we tied ourselves even more tightly to Trump and looked away. As I look back, it felt like a classic abuse relationship—something that I unfortunately know a lot about from experiences in my childhood and some romantic relationships throughout adulthood. I won't get into details, but suffice it to say that I had become an expert at making the best of a bad situation. The secrets you hold. The lies you tell yourself. The ways to keep your abuser from becoming upset with you. The fear, the panic that come from anticipating who he might be on any given day or at any given moment.

At the White House, Trump was the distant, erratic father we all wanted to please. I tended to forgive his sins, forget his foibles, believe that he was better than outsiders were saying he was. When he liked you, when he was pleased with you, he overwhelmed you with charm and generosity and even affection. And when something set him off or someone else did, he'd start screaming. His temper was terrifying. And it could be directed at anyone, whether he or she deserved it or not.

When I saw my phone ring and I knew it was him, I'd feel a panic in my chest. Would he be happy? Would he be mad? What did I just do (or not do) that might have pissed him off? How can I make him feel good again? I had the news on 24/7 (CNN, much to everyone's irritation), but still when he called I would scroll Twitter frantically before I answered. I had a notebook with me at all times so I could tell him "how things were playing," ask him questions, or transcribe things he'd want me to say.

I was wrapped up in this cycle for years, and to be honest, I had few options anyway. I was a single mom with no trust fund. If I had quit earlier, where would I have gone? Corporate America was not going to welcome someone from the Trump White House with open arms. The Trumps were all I had. At least that was what I believed for a long time. So I stayed and endured and tried to make the best of it. Many of us did. And we saw what happened.

AS I WRITE THIS in the early days of 2021, we have a new president and a new administration. I have been commuting between DC and Kansas for many months now. After my resignation on January 6, I was summarily dismissed by the Trump family and many friends and colleagues in a way that my ego had never allowed me to think could ever happen. To this day I have not heard from a majority of the people I worked alongside for years and years, all for doing the one thing Mrs. Trump had instilled in me constantly and that she told women and children all over the world to do: "Always stay true to yourself."

In the past months, I have been able to reflect on the past six years: why I joined the campaign, why I entered the White House and stayed for so long, what I saw, and, most important, what I learned. This book, which started as a personal journal, is not just about politics or the White House. It is about loyalty and family, learning and really fucking up, proud moments and monumental regrets, narcissism and humility, love and heartbreak, friendships and loss, and of course falling down and trying your damndest to get back up. As someone once said, "Stockholm Syndrome is what it is when you begin to identify with your captors—they get nicer every day that they don't kill you." It is a story that has not been easy for me to tell. There were many times in writing this book when I had to take a moment or a day and focus on something else. Emotions surfaced, and I forced myself to relive some painful experiences.

I realize now that I had devoted much of my time over the past six years to people who in the end probably didn't deserve my taking time away from my friends, my family, and my two boys, whom I don't intend to mention again in this book. One of the things I learned from Mrs. Trump is that kids deserve privacy, and that includes my own. I became excellent at compartmentalizing that part of my life. Looking back, it was not only to protect them, but I think to block the pain of how much I missed them every day and the uncertainty of whether I was doing the right thing.

This is not a book, by the way, where you need to like me. I am not trying to win people over or gain moral absolution. But I do think this is something people need to read because I observed a truly unique, scary, bizarre, often funny, riotous, wild, and at times tragic period in our country's history. I want to answer all the questions I think I would have about the Trump White House had I not been there: What were they really like? Why did people put up with things? What was that marriage really like? What was the deal with the Russians? Was it all as wild and crazy as it seemed?

That era will be talked about and remarked on long after I am gone. I saw a lot of it from a unique vantage point, working simultaneously for both the president and first lady. I was their adviser, their underling, their annoying nag, their gossip buddy, even sometimes their friend—or so I told myself. I liked them and I disliked them and I miss them and I hope I never see them again.

I

We Won—Now What?

Have the humility to learn from those around you.

—JOHN C. MAXWELL

I first officially met Donald Trump in a bathroom, only seconds away from disaster.

I was at the Iowa State Fair in the months ahead of the 2016 Iowa caucuses. We had been flying, then driving around the state for quite a while, I'd been drinking coffee after coffee to keep alert, and, well, I needed a bathroom ASAP. As soon as we arrived at the fairgrounds, I ran to the management office and the people there directed me down to the basement, where I found a very tiny bathroom with a sink, a toilet, and a door with no lock.

As I was finishing up, I could hear a commotion outside but thought little of it as the enormous state fair was under way just above me. But as I was washing my hands, the noises got louder and suddenly the door swung open, bringing me face-to-face with Donald Trump. Just me, him, two security guys, and a toilet. The space was so tiny that his head almost touched the ceiling—I had been volunteering for him for a couple of months but hadn't realized how tall he was until that moment.

All I could think was "Thank God he didn't arrive two minutes earlier." I knew of his legendary germophobia and suppressed a childlike urge to let him know I had only gone "number one." Instead I just stood there, frozen in his presence. Maybe he was used to that. He finally motioned for me to step out past him.

Then, with a spark in his eye and a kind smile, he said, "Look, I'm not too proud to use this bathroom even if it is a ladies' room. It can be our secret." Instead of laughing or responding or acknowledging that he had broken the ice in a gracious way, I ran past him without a word. So although my perception of him that day was that he had made an awkward moment bearable, I'm sure he thought I was a complete idiot.

Thankfully, as time went on, he started to recognize me as the girl who was always with the press corps. I'd had the same role as a press wrangler with Mitt Romney and Paul Ryan in 2012, then worked as a press secretary for the Arizona attorney general, followed by the Speaker of the House. I had also held communications positions with some PR firms, so I was well versed in all things press/communications and media relations. I fervently hoped he had forgotten all about our weird bathroom encounter, but he never mentioned it and he was not one to keep passing thoughts in his head to himself. What he did say on a regular basis was that he appreciated the way I "handled" reporters. He often commented to me that he was surprised that they actually listened to me and we all seemed to get along. He clearly didn't realize that I sometimes spent up to twenty hours a day with the traveling press corps, and they knew that if they fought me too hard, the days could turn quite unpleasant. I think Trump always wanted a better relationship with the press but didn't know how to get it, as he had done so easily in his *Apprentice* days. More on that in a bit.

*   *   *

ELECTION NIGHT 2016 WAS a blur. I was working and with the press pool as usual. I was happy to have some of my "originals" in the pool that night—John Santucci with ABC, Noah Gray and Jeremy Diamond with CNN, Ali Vitali with NBC, and Sopan Deb with the *New York Times*, to name just a few. Because no one actually thought that Trump would beat Hillary Clinton, the veteran, more experienced reporters had all clamored to be assigned to her, leaving openings for other reporters to follow the Trump circus, where they ended up becoming household names, such as Jim Acosta, Hallie Jackson, and Katy Tur. I got along with most of them very well and came to like many of them. Because nothing in the Trump campaign ever resembled an ordinary campaign or even made sense most of the time, let alone followed a clear plan or had a coherent message or strategy, we were kind of all thrown into the logistical shit show together, feeling our way in the dark. On election night, we all sat in the small ballroom at the New York Hilton Midtown for hours, many in the room waiting for the race to be called for the Democrats. Yet that never happened. It seemed that with every state we won, the air in the room got easier to breathe. I spent most of the evening in the buffer with my reporter and photographer buddies while all my campaign colleagues got drunk off both their happiness and the abundant supply of alcohol all around us. The buffer is the space between the stage and the people in the audience, built for safety reasons but also allowing a small group of roughly twelve to fifteen members of the media to be up close and personal with the candidate or, in this case, the new president-elect. In the wee hours of the morning, President-elect Trump and the family came out.

As he walked toward the podium, he pointed at me as he always did, having come to recognize that I was the person on his team who was always with the press. I don't generally cry, but I did that night. It was the culmination of long hours, days, and months. Of being away from my home in Arizona, my family, my best friend, and others. It was the realization that the chance I had taken had actually been worth it and all the people—family and friends alike—who had turned their backs on me would have to at least acknowledge that it wasn't just me—that half the people in the country had chosen this man to lead the free world.

AS A RESULT OF my job as liaison to the press corps, I had a unique seat at many meetings in the early days of the Trump transition. I was the one who usually brought the press in, and I stood less than ten feet away from Trump while he performed for them. I choose the word "performed" deliberately because sometimes he seemed as though he was still hosting a TV show. It got to a point that he would seek out my face as soon as I brought the reporters in. I liked to think it was because I was probably one of the only friendly faces in that crew. Whatever Donald Trump said about the media publicly, I think it bothered him, especially early on, that he didn't have the rapport with or respect from them that he'd had when he was an entertainment figure. He had said nutty, batshit things for decades, after all, and reporters had laughed or brushed them aside. Not anymore.

Trump was a hungry gossip. He consumed information about people almost as eagerly as he consumed Diet Cokes. Many times after I dismissed the press from a photo op, he would motion for me to stay back so he could ask which reporter had been the nicest, which had given me problems, who was friends with whom, and of course what they had privately said about him. Or sometimes if a reporter asked what he perceived to be a "nasty" question, he would ask me what I thought about the person and why we had allowed him or her in. That became quite a needle to thread because I respected the press corps and the job they did, but I also knew that if I told him some of the things they had said, he would either repeat it back to them—he has no filter, as I learned quickly—or kick them off the plane or out of the White House. Most of them clearly leaned left politically and seemed almost to consider the fact that Donald Trump was now president a bizarre joke. No good would come from my telling him their thoughts or conversations, so for the most part I came up with a standard response. I would generally say that a certain reporter was being "a bit difficult today," roll my eyes, but then add that he or she was in awe of how far Trump had gotten and all that he had accomplished. That seemed to keep him happy, knowing I wasn't best friends with any of the reporters while also stroking his ego. I don't know if I was consciously trying to advance myself in his eyes that way, but it certainly didn't hurt. I think perhaps he came to think of me as his spy.

My role became especially important to him during the transition, when he brought candidates for cabinet posts to his golf club in Bedminster, New Jersey, and trotted them out in front of the cameras. He placed great importance on making a "moment" or an event of the interviews, and he would always ask me what the press thought of each candidate. Did they think Chris Christie was going to get a job? What did they think of Rex Tillerson? Afterward, Trump would arrange for the cameras to capture both him and the candidate as he said goodbye to them. Those were often times when Trump would say a few words to the press. It was a master class in television production, and he was a natural producer. It was also the world he knew best, a sort of *Celebrity Apprentice* situation but for actual cabinet secretaries, and the press ate it up. Fundamentally, Trump wanted to impress reporters.

Maybe, he seemed to think, they would start to write good things about him again.

It was a little petty of me, but my favorite of these interviews at the time was the one with Mitt Romney, who at least some in the press believed was a serious candidate to serve as Trump's secretary of state. Romney apparently thought so too. The Trump-Romney feud, of course, had been very public and bitter. During the 2016 race, Romney had memorably denounced Trump in some of the harshest language the genteel former governor ever used. He'd said that Trump "lacks the temperament to be president" and that "dishonesty is Donald Trump's hallmark." He'd called Trump "a con man" and "a fake." And he, like many other #NeverTrumpers in the Republican Party, had condemned people who supported him or worked for him. Trump had responded as he always did with his usual bazooka blasts of insults: Romney was a "loser" and a "failure" and so on, including one of his most random, "Romney walks like a penguin." Could Mitt Romney forgive and forget now that Trump was elected? Apparently. Could Donald Trump? Well, read on.

I had worked for the 2012 Romney campaign as an advance person. Though I'd spent limited time with the candidate, both he and Paul Ryan were always kind and appreciative of the people who worked for them. When I began working full-time for Trump, I received a number of messages from my 2012 colleagues telling me that I was ruining my career and how disappointed they were with me. So I'll admit that I felt an odd sense of childish satisfaction when Romney, of all people, was suddenly on the list as a potential secretary of state and appeared to be gunning for the job.

After Romney's first dog-and-pony-show meeting at Bedminster, there was much speculation among the press corps that President-elect Trump was meeting with Romney only to gloat and as payback for all of the criticism Romney had lobbed at him just weeks earlier. I didn't believe that at the time and thought they were just stirring up trouble. Why would a newly elected president waste his time on something like that? He had won; there was no reason to rub it in. But I learned very quickly that that was *exactly* what Trump was doing, a revelation that was both surprising and not surprising at the same time. *Of course this is something Donald Trump would do.* The whole ploy was an open secret among key Trump advisers such as Jason Miller and Steve Bannon, who were delighted at Romney's willingness to put himself through the humiliation, eat his words from the campaign, and still not get the job. That was my first lesson that these guys played for keeps.

One day when we were back in New York City, headquartered at Trump Tower, where the president-elect stayed until he moved to the White House, Jason Miller called to ask me to have the press ready that evening for an off-the-record stop at a restaurant in town. In political parlance, an off-the-record movement meant that I could tell the members of the press where we would be going but they could not report it until we arrived at the location. This is most generally for security purposes but also so that the site doesn't become crowded with fans and protestors who would hear hours earlier that Trump was planning to show up.

"What's going on?" I asked Jason. Herding the press corps around with few details wasn't easy, and I knew I'd get endless questions (and complaints) from reporters.

He told me that there would be a dinner at Jean-Georges restaurant in the Trump International Hotel & Tower in Midtown. I needed to be ready to bring the press into the dining room to get a shot. He then revealed who was going to be in attendance: Trump and Mitt Romney.

Though he didn't say it in so many words, Jason made it obvious that Trump had cooked up the dinner, so to speak, just to torture the guy a little more. The setting was Trump's turf—a restaurant in his building and one of the best in the city. The place had three Michelin stars and four stars from the *New York Times* (probably the last time they would be so nice to a Trump property) and was run by the world-renowned chef Jean-Georges Vongerichten. But the point wasn't to treat Romney to the best, it was to show him who was in charge. Trump wanted all the press to see that Romney would come all the way to New York and sit down with a man he had called a "con artist" and "a fake" to sing for his supper. Donald Trump was many things, but even his critics had to admit that he was a master at TV spectacles. This was yet another, set to be one for the ages.

Several members of the team, including Dan Scavino, Trump's social media guru, and campaign adviser Jason Miller, were especially proud of that photo op. But when Trump's designated White House chief of staff Reince Priebus ended up attending the dinner, too, I thought the whole affair might turn out to be more civil than what I had originally prepared for. I had been imagining something like Trump shouting "You're fired!" right after he finished his steak and then watching Romney slink out of the restaurant "like a penguin."

As in most restaurants in New York City, the space was small and the dining room was packed, with little distance between tables. This presents quite a logistical challenge when you have twenty members of the press, some with TV cameras and boom mics, the rest with cell phones and mini recorders, all trying to get as close as possible to record the scene. Based on the sneers and dirty looks—welcome to New York!—the rest of the patrons in the restaurant were not at all pleased, nor were the waitstaff for that matter. They had to juggle serving trays and plates around an angry crowd of regulars, a VIP table with the president-elect,

Secret Service agents, and an aggressive, impatient press contin-gent. We stayed in the dark restaurant for a good ten minutes, flashing lights, shouting questions, hitting people in the head with equipment or backpacks. Everyone was stressed and annoyed and hot and miserable. Well, almost everyone. One person in the room wasn't annoyed, pissed off, or even slightly bothered. The president-elect loved it all. His seat at the table was facing the press, and he made eye contact with every one of them, flashing big smiles and a thumbs-up. He was basking in his glory, the cen-ter of attention, center of politics, center of the world. And some poor sap beside him was about to be his main course. The dinner began, by the way, with frog leg soup. I'm not sure whose idea that was. Maybe Trump ordered it because he figured that was something a fancy guy like Romney, whose wife taught horses to dance, would want to eat? Maybe he wanted to show off a Jean-Georges specialty? Who knows? But knowing that Trump's own dining habits resembled those of a sixteen-year-old, I couldn't see it being his idea.

The press got their photos of the three of them at their table, and it would be one of the enduring images of the Trump era. The shot everybody remembers focused on Trump and Romney, with Priebus out of frame completely (an appropriate metaphor for his entire White House tenure). Trump is flashing a huge grin, while next to him, Romney looks awkward and uncomfortable—like someone who had swallowed a penguin rather than walked like one.

As I watched the scene unfold, my earlier petty feelings disap-peared in an instant. I felt sorry for Romney as he sat there like an animal in a zoo, glancing around nervously and politely, clearly uncomfortable with the spectacle, as he prepared to receive his just desserts. It was cruel. If anyone else on the Trump team felt that way or even had a shred of regret about the scene, they kept it secret. Somewhere inside me a little voice told me that it was all so wrong, but then the voice was gone. I had a job to do and I was going to do it, even if I didn't think this was the right way to go about it.

After the photo op, I was told to hold the reporters in a room until the president-elect left, then move them to the sidewalk. That, too, was for a purpose: so the cameras could capture Governor Romney walking out all alone, looking pathetic. I assumed it meant that Trump had told him at dinner that he wasn't getting the job and they were leaving it for Romney to tell the assembled press corps he was not selected. At least, I thought, this will be over now. Instead, as he stood there alone, left on the side of the road, Romney said he was "still hopeful" about getting the job and that the conversation with Trump had been "a good one." That was one of the first times I learned that Trump never liked to deliver bad news in person. Instead he'd let Romney's agony continue a little longer. Nearly two more weeks passed before Romney announced on Facebook that he wouldn't be serving as secretary of state, while adding that he was "hopeful" about the Trump administration. I don't know how aware he was at that point about Trump's deliberate, slow torture of him or if he was a particularly vengeful man. But in the years to come he would find ways to get back at Trump.

AS USUAL, I WAS forgotten over the next few days. Since I was always stationed with the press, no one higher up on the Trump team really knew what it was I did or all that went into the logistics of organizing a press pool. With the exception of George Gigicos and Kellyanne Conway, who had worked on campaigns before and understood my role, I was known simply as "the press girl." Now that Trump was president-elect, my job would be taken to a more official level, but because this was all new to pretty much everyone on the team, no one realized or had the time to think about the fact that someone would still need to organize or escort the protective pool if the president-elect made a movement. Whereas most everyone else on the team was named to the transition team or went back home to prepare to move to DC and work in the administration, I stayed in New York, not knowing what to do and getting no direction. It was another lesson of Trump World: just do what you want to and hope it works out. So that was what I did. If I heard that the president-elect was going somewhere, I mobilized everyone in the press to get into the motorcade. No one told me to do it, I just did it—except for one time in November when he and his family took off for a steak dinner at New York's famous '21' Club without telling anyone. The press were none too pleased, but we caught up to them and stationed ourselves inside a taco restaurant across the street, but, man, that was a shitty night.

In the meantime, I had a future to figure out, and eventually things started to come together. George Gigicos, who was the head of the advance operation for the campaign, let me know that he had accepted that same role in the White House and offered me the job of deputy director of press operations. Basically, it was the job I was already doing but in the White House. George was the one who had "brought me to the dance" and onto the campaign in the first place, and he's one of the best in the business. We had both worked advance on the Romney campaign and had become good friends, so I was excited at the prospect of working with him in the White House. I quickly said yes.

I rented an apartment sight unseen in Washington, DC, and packed up the random boxes of clothes and items I had collected in my travels over the past year and a half. I moved down when the president-elect did, just a few days before the inauguration.

For a while I slept on an air mattress and "borrowed" disposable coffee cups from my new lobby.

I discovered a nice tradition in Washington: the outgoing party in the White House show the ropes to the incoming officials, regardless of their political party. Barack and Michelle Obama graciously welcomed the Trumps to the White House after they won, and various Obama employees did the same for the Trump people assuming their jobs. None of them was happy about Trump winning, and some of them were more sour about it than others, but still it was a courteous thing for many of them to do. I met with one of Obama's press wranglers, who showed me around the White House to explain how things worked. He was even kind enough to take me to the colonnade and point out President Obama working in the Oval Office. That was an incredible experience in so many ways, and I knew it had to be tough for the departing team on many levels.

A FEW DAYS BEFORE the inauguration, I got a call from Sean Spicer, who had been named the first press secretary in the Trump administration. I'd liked Sean from the first time I'd met him. He asked questions about me and solicited my thoughts on various press matters, and unlike many people in politics, he seemed to genuinely want to know the answers. He could be fun, too, with a strong sense of humor and an eccentric habit of chewing gum and then swallowing it. He chewed and swallowed more than a pack a day.

Spicer was Reince Priebus's "guy" and clearly knew his shit in terms of Republican politics and communications. The flip side of that, though, was that he knew his shit because he'd been around a long time. He was an "establishment" figure, having worked in DC since the 1990s, including stints as a Capitol Hill staffer and

at the Republican National Committee. When certain Trump allies grumbled about having too many "RNC people" moving into
the White House (especially because the RNC had basically left
Trump for dead after the *Access Hollywood* scandal), they meant
people like Sean.

I sensed that Sean might be "too traditional" for the boss and
wondered how the rest of the Trump crew would react to the
"new person" speaking for the new president. Donald Trump was
his own spokesperson, and I didn't imagine that he would adjust
well to the idea of someone else, someone he didn't know or necessarily trust, having the temerity to speak on his behalf.

Sean offered me the job of deputy press secretary. Lindsay Walters from the RNC was also given that title, and Sarah Huckabee
Sanders was named principal deputy press secretary. For me, the
job offer was truly a dream come true. For as long as I could remember, I'd wanted to be White House press secretary "when I
grow up." I'd watched so many people on TV do the job well, such
as Dana Perino and Ari Fleischer. I imagined myself standing behind the podium, with its presidential seal, and filling their shoes
on matters of national and even international scale. I wasn't close
with many of the RNC people, so I was surprised and honored
when I got the call that would place me on a path to my dream
job. I accepted Sean's offer immediately, then went to the difficult
task of calling George to tell him that, uh, on second thought,
I wasn't going to do the other thing after all. I felt guilty, and I
feared he would be angry. To the contrary, he told me that I'd be
stupid not to take the job, and we discussed some good options for
my replacement.

As soon as I hung up, my heart sank. I remembered the two
DUIs I had received back in Arizona. Of course, I had noted them
on my background and security clearance forms, along with an
explanation of each, but I knew that as deputy press secretary I

would be taking on a much more public role in the administration. And I figured that nobody would care about the backstory. Reporters and Democrats would not care that five years prior, I had not noticed the speed limit change from forty-five to thirty-five miles per hour and been pulled over and admitted to having drunk two glasses of wine, which had resulted in a reckless driving charge. They wouldn't care that Arizona has the strictest DUI laws in the country (which I have no objection to). And certainly no one would believe the story of the second incident, which was that I had been at a Christmas party for work and was moving my car to a parking lot so someone else could drive me home. In the time it took to back out of the street parking and drive across the street, a policeman noticed that my headlights weren't on and pulled me over—ironically in the parking lot I had intended to keep my car in for the night. The coworker who was following me and planning to give me the ride home even walked over and informed the policeman of our plan, but again, Arizona has a no-tolerance policy and I was charged.

I had already been so ashamed and so disappointed in myself for the past five years, and as I prepared to call Sean, the moment only drove home further my anger at myself. I had grown up with some family members who had issues with drugs and alcohol and gone to great pains all my life never to let that become who I was or was perceived as. I knew that there was no way I could represent the president of the United States with that on my record, no matter what the context had been. I didn't want to be a person who reflected poorly on the new administration, so I picked up the phone, called Spicer back, and told him everything. My dream job, I figured, would evaporate just as quickly as it had appeared.

To say that Spicer was gracious was an understatement. He said as long as I had been honest on my background check information, the records of the arrest would all reflect the circumstances and it would probably be okay. I will never forget that night—and how understanding both George and Sean were for vastly different reasons. I will also never forget the sequence of going from thrilled to guilty, from over-the-moon happy to terrified and disappointed, to finally just hopeful and grateful for the chance I'd been given. As it turned out, it was great practice for the gamut of emotions I would run over the next four years.

2

"This Is Going to Be a Shit Show"

Chaos, when left alone, tends to multiply.
—STEPHEN HAWKING

My first day in the White House was a blur, just as my visit there during the last days of the Obama administration had been. I tried so hard to remember it, to savor every moment, yet I had already forgotten most of what I had been told that day. I think that tends to happen the first time you are inside the walls of the West Wing and are shown things most people never get to see. I was walking through history, and it was challenging to keep my bearings.

Most of the communications team was placed in the Eisenhower Executive Office Building, a massive gray French-style imperial building across the driveway from the White House. It was originally built in the 1870s and 1880s and known first as the State, War, and Navy Building, as all three departments were housed there. These days, it holds probably 90 percent of all White House employees.

It's gigantic. The black-and-white-checkered marble floors distracted me, I was confident I'd fall down the steep staircases eventually (I never did), and there were all these miniature, short doors—to this day I don't know what was behind them or why they were so small. But the building had so many wonderful details. Some doorknobs, for instance, have eagles, swords, or anchors on them, depending on whether the door is in the original State, War, or Navy section of the building.

We were all placed in different lines to receive briefings, security badges, access badges, laptops, and cell phones and be fingerprinted. It was fascinating to watch and be part of the process that was the White House Military Office onboarding hundreds of people in order to keep the government running.

Those of us with offices in the West Wing had a similar experience. More than once I found myself angry that the creators of the TV series *The West Wing* hadn't done better research. As an avid watcher of the show, I blamed them for my unrealistic expectations. All of us got lost constantly; trying to find power outlets and light switches was comical; the bathrooms were a mystery to me, and no matter how many times I stumbled upon them, I could not remember how to find my way back. There was a cafeteria/convenience store in the EEOB that served hot food and snacks, and in the bottom of the West Wing was the Navy Mess, but none of us learned about those important survival spots for probably a week.

A big highlight of day one was the folder that the White House Historical Association leaves on your desk containing a welcome letter and a book called *The White House: An Historic Guide*, meant to provide historical context to all new staff, including a history of the various rooms and a list of the people who occupied the office before you.

On day one we also found some notes (and some bottles of Russian vodka, along with a cabinet full of books written by Barack Obama) left over from someone in the previous administration, or so we assumed. One note read, "This is going to be a shit show."

At the time, it pissed me off, but I also figured that it was a thing that happened when administrations changed out, so I didn't give it much thought.

A couple of years later, reporters didn't believe me when I told them that the Obama people had left those "gifts" for us when they had left. Susan Rice, Obama's national security advisor, called it "another bald-faced lie." Another Obama national security aide called it "shameless and disgusting" and said that I, specifically, needed to be fired.

But years later, my story was mostly validated by an extremely unlikely source: the actor and comedian Dave Chappelle. He admitted that the notes were actually real but that it wasn't staffers who had left them (which I had thought was a reasonable assumption at the time). In an interview on Naomi Campbell's YouTube show in April 2021, he recounted that during a farewell party at the Obama White House featuring a bunch of celebrities, it was the boldface names—which he refused to divulge—who had written the notes and hidden them. "I saw this happening. I'm not going to say who did it," he said. "But it was celebrities writing all this crazy shit and putting them all over there. I saw them doing it, so when I saw it on the news, I laughed real hard."

At least he had a better sense of humor than Susan Rice did. But to whichever Hollywood liberal wrote the note about its being a "shit show"—you may have been trying to be nasty, but it turned out that your note was prescient.

SEAN HELD HIS FIRST briefing in the early afternoon of January 21. As with everything we did, it was pretty much chaos. The press briefing room, which is smaller than it looks on television, was packed. The "mainstream" reporters were in their assigned seats, and there were many other reporters crowded along the sides and in the back. Photographers were crawling over everyone else to get the first shots. I was seated at the side of the room closest to the press, so I had the good fortune of being knocked in the head by huge still cameras because once photographers get behind that lens, they see nothing but the subject they're shooting. Although she wasn't on the communications team, Omarosa Manigault of *The Apprentice* fame was to my left. Like a supernova, she would burn out quickly and then turn on the president in spectacular fashion. I liked Omarosa on one level, as she was always friendly and easy to talk to. On the other hand, she was a reality star with a tendency toward drama. She allegedly misused the White House car service, or so I heard, by having it drive her to doctor's appointments. I was also told she tried to have her wedding photos taken at the White House, and when she didn't get permission from the Office of the First Lady, she showed up at the gates with her wedding party and made a big scene anyway. I didn't personally see that happen, but it sounded about right. It was almost always a big scene with Omarosa, which of course was what had made her a TV star.

Hope Hicks was next to her, then Kellyanne Conway, and finally Sarah Sanders at the other end. I remember being proud that the press team supporting Sean that day consisted of all women. I had no idea what Spicer planned to say, but the briefing was being held on a Saturday and I remember thinking that that was odd. My hope was that it would just be a fun and introductory kind of briefing, but when Sean didn't appear and the minutes started to tick by, I sensed trouble. The president was already angry about press reports saying that his crowd size at the inauguration was smaller than Obama's. By that time I had been around long enough to know some of what triggered the president, and any unfavorable comparison to Obama, especially a suggestion that Trump was less popular, was at the top of the trigger list. I couldn't help but feel nervous for Sean. He was in a no-win position because there were no magic words to persuade people that Donald Trump's crowd was bigger than Obama's.

Everyone now remembers the press briefing that ensued— when Sean insisted that Trump had had the largest inauguration crowd in history, despite what we had all seen and knew. Sean's credibility with some of the media and the public was shot from then on, and there was nothing he could do about it because he had said what the president told him to, and that was how he saw his job. The press secretary serves at the pleasure of the president, and Donald Trump was not pleased.

I won't dwell on that disaster any longer except to say this in Sean's defense: I ended up having the same job, as Trump's press secretary, so I know firsthand how much pressure he placed on you. He watched the briefings, commented on everything from what you wore and what you said to what reporter you took a question from and what reporter you ignored. He expected and demanded that you say certain things regardless of whether they were true or not. Sometimes he dictated things word for word for you to repeat. Forcing Sean to claim that the inauguration crowd was bigger than Obama's, which I imagine Sean also knew was bullshit, was a test. Trump always wanted to see how far you would go to do his bidding; it was his way of measuring your loyalty. And it was hard when you were caught in the middle of something like that to keep your bearings. I remembered that lesson when I took over as press secretary and resolved to do whatever I could to avoid having Trump do to me what he'd done to Sean. I wasn't always successful, but I tried.

SEAN WAS ALSO PARTICULARLY vulnerable because he was seen by some as an outsider. In those first weeks, the internal gossip that people were reading in the press was true. It was very much an "RNC people versus Trump people" atmosphere. I don't think that was anyone's fault; it was just two very different groups and sets of agendas coming together not just to work together but to run the federal government. The Trump campaign was small and had always been independent and scrappy. A big part of why he won was his commitment to cutting through bureaucracy, running his affairs like a successful business, and bucking tradition. The only problem with that plan was exactly what made the campaign so great: it was too small, and it broke too many norms. Once we got into office, we were overwhelmed by the size and scope of the federal government and how many professionals it takes to run it. There were dozens of agencies with thousands of employees working not only to implement new policies but to also ensure that current policies were being carried out. I don't think many of the senior officials had a clue about any of that, and that really hurt the administration from the start. Because we had alienated so many of the "normal" Republicans in Washington, a lot of them were either unwilling to join the administration or were put on some do-not-hire list by someone in Trump World.

The president named Reince Priebus, who had been serving as the chairman of the Republican National Committee, as his chief of staff. Although Steve Bannon had been given an equal amount of authority, many on the "Trump side" were still not happy about Reince's being given such a powerful voice and felt that Bannon was the default chief for the Trumpers. The fear was that Reince would treat his RNC people better than the Trump people where jobs and attention were concerned, which he did. Looking back now, it makes sense that he would favor those he had existing relationships with, but the *Hunger Games* vibe that endured throughout the next four years had taken root.

The word on the street quickly spread in the Trump White

House that face time with the principals was everything. If you weren't in the president's line of sight, he'd forget you. As a result, Sean guarded his access to Trump and dominated most televised interactions. In my first two months as deputy press secretary, I didn't do any interviews but was relegated to my old role as the wrangler for all of the press sprays (the moments when reporters and photographers were herded into the Oval Office to take photos or try to ask a question or two or twenty).

That actually turned out to be lucky for me. I didn't know it yet, but there were very few people Donald Trump was always happy with. I was in that category. Since I wasn't Spicer, the president didn't blame me for the bad press coverage he was getting. Instead I was the person who brought him what he loved several times a day—attention from the media—and I was the one who ushered the reporters out when he was tired of them. The president liked what he thought was my "tough" style of moving the press out of the room when it was time to go. In his vocabulary, there was no higher compliment than being "tough" or "vicious" or "a killer." I think that was why Sean kept me in the wrangler role long after I had outgrown it. Keeping the president happy was all that mattered.

It also gave me a front-row seat to the action. Seeing who was going into the Oval Office and who wasn't made it easy to identify who the power players were. The dynamic of Bannon and Priebus made for a weird and entertaining spectacle. They could not be two more different guys—the Felix and Oscar of the early Trump White House. Reince was always composed and well dressed, albeit a bit awkward. Bannon—well, he was something else, with the multiple shirts-over-shirts combos that he favored and his wild, stringy hair. I was shocked that any senior official in the White House could dress so slovenly and even more that the president allowed it. Trump himself was almost always in a suit and tie—those ties, how he treasured them.

Usually Bannon would lock himself in his office with the white-board that he seemed obsessed with and scribbled stuff on, plotting who knew what. He came into office with grand pronouncements and a plan to remodel America in Trump's likeness. But Bannon, who was always friendly to me, maybe even a little shy, was gone in about a year.

With Priebus, I saw a pattern that would play out with every chief of staff who followed him. (There was a total of four in four years.) First, Trump would announce the chief with great fanfare, promising him carte blanche to manage the White House as he saw fit. "Hire who you want. Fire who you want. You are in charge." That sort of thing. That was the honeymoon phase. The rest of us would give the new guy respect and attention. Even Jared Kushner (the real chief of staff) would play along for a little bit. That would last a couple of weeks. Then suddenly Trump would get tired of the new guy's rules and empower other people to work around the chief, which they did. After that, the chief would get bitter, become disillusioned, and then be gone, and we'd have a new one in place. The fanfare would start all over again. The same was true of the cabinet. After a while I stopped paying attention to them at all. They were supporting players, at best. There was that guy Rex Tillerson and then General Jim Mattis and then they were gone and then someone else came in and then another guy. I'd actually have to look at a list to remember who they all were—a blur of mostly middle-aged white dudes whose ultimate influence proved to be fleeting, to be honest. That was Trump's cabinet.

BUT FOR ALL OF the press grumbling that the Trump White House was a game of musical chairs, although that certainly applied to the cabinet, there actually was a core group of people Trump relied on who stayed pretty much the whole time. Among that enduring group of loyalists were Peter Navarro, who was an adviser on trade policy but saw himself mostly as Trump's henchman and loyalty enforcer; Stephen Miller, the president's chief speechwriter, who was more than the caricature that some in the media, and sometimes he himself, created; Larry Kudlow, who was the Director of the National Economic Council; and Dan Scavino, the president's director of social media and arguably the most important person in the entire White House because he commanded Trump's most prized possession: his Twitter account. In the communications shop, where I worked, Hope Hicks and Kellyanne Conway were two key advisers who also stuck around for most of the Trump presidency.

I had heard the name Hope Hicks before I met her. The first time I saw her, she was wearing a stunning bright orange dress. I was flabbergasted because I had somehow expected a no-nonsense middle-aged woman in a pantsuit. Hope, Melania, Ivanka, Lara, and Tiffany Trump, Omarosa. How many fit, beautiful women were in this world? Hope was always friendly and polite, but I was never her priority. She never forgot who mattered the most to her: the president, Jared, and Ivanka. That was it, and I do not fault her for it. She focused on them almost exclusively, leaving her little time or interest in the rest of us. I worked very hard for her approval at first; then I tried not to care that my efforts never worked. Finally I recognized and respected her no-bullshit attitude. She didn't need to waste time being fake to people, and she made no apologies for it. Hope was very savvy and smart, and the president trusted her advice implicitly. She also had a way of calming him down and slowing down some of his ideas when no one else could, which was a huge asset when it came to Donald Trump.

Kellyanne Conway is one of the strongest and most cunning people I know. Words are her weapons, and she is a master at the art of verbal gunslinging. That woman could have a conversation with you, make you feel empowered and in control, then throw a barb or two in there that you didn't even internalize until you'd left her office after giving her a hug. She can cut you down so painlessly that she's already down the hall in another conversation by the time you realize you are bleeding. She is also no bullshit. She fought like hell for junior staff, and for me when the going got really tough. She was also passionate about topics such as health care, opioid addiction, women's issues, and the dangers of e-cigarettes and was one of the president's staunchest defenders on TV. Over the years I watched Kellyanne field incessant questions about her marriage from reporters, deal with a spouse who was none too shy about his thoughts about who we worked for, and never blink an eye. I marveled at how she managed to keep her composure when her own husband was vocally supporting her boss's impeachment and removal from office. Kellyanne could give tough love and perspectives that you didn't want to hear but needed to. Like so many of us, there are layers to her: the TV warrior, the first winning female presidential campaign manager, a woman in a complicated marriage, a mother, a friend, and a professional. Like Hope, the president appreciated her perspective and kept her around despite what seemed the best efforts of both Jared and Ivanka, who appeared certain she was a prolific leaker.

Then, of course, there were the two who would never leave: Jared and Ivanka.

One of the unexpected developments during my time in the White House press office was the role we were expected to play with the first daughter and her husband.

I didn't leave the White House fans of either one of them, as will become clear. But my first impressions of both were positive, maybe naively so. The minute I met Ivanka and Jared, I wanted them to like me. They seemed so glamorous and well put together, in their trim, form-fitting outfits and their perfect teeth and hair, while at the same time seeming down to earth and non-judgmental. I marveled at how Ivanka could look so effortlessly spectacular early in the morning and stay that way all day long. She was well spoken and commanded the room.

Jared Kushner was handsome and charming, calm and casual. He could be funny, was clearly smart, and was kind to every junior staffer in the building, remembering their names and small details about them. He was skilled at making people feel important and as if their ideas mattered. In a White House that so often cared only about senior staff, I admired how attentive Jared was to people who only saw him in the halls on occasion. His voice was very quiet and his cadence so even that I was caught off guard whenever he dropped an F-bomb, which he did on more than one occasion. Anyone with a brain in the White House instantly knew that Jared and Ivanka were the real power players, the people to be reckoned with, the ones you didn't want to cross.

From the very start, Ivanka made it clear that she expected the White House press office to siphon off some of its resources to defend and support her. She obviously had a Google alert set for her name and would go to Sean Spicer whenever a story about her popped up that she didn't like, which was most of them, expecting us to push back. That happened even if 90 percent of a story was positive. She would focus on obscure small facts that she didn't like or claimed weren't true.

One story that particularly upset her was the rumor that as a teenager attending Manhattan's elite Chapin School, she and some friends had flashed a sidewalk hot dog vendor from the window of their classroom. To me, that factoid was pretty harmless. Like many women, I had my own wild side on occasion and, like every young girl, did stupid things in my day. Responding to something like that would only amplify its importance and give more oxygen to the story. But Ivanka didn't see it that way. Image was everything in the Trump family, and Ivanka worked very hard to convey an image of perfection, with her handsome, thin husband (I think I saw him eat a large salad once) and her truly beautiful children. Flashing her assets? Well, that didn't play in that narrative. She demanded a forceful response, as she usually did, and so Sean accommodated her as best he could.

OF COURSE, THE PRESS coverage of the Trump White House was from start to finish disastrous. Some of that was our own fault, but the press shares in some of the blame, too, as a number of stories written in the thick of the Trump-media war have since been proven wrong. Because many of us on the comms team had never done this before, we gave out contradictory information or no information. Or we'd say something and then Trump would undercut it. Some of it was inevitable and deserved. But sometimes it felt as though we could never do anything right, with reporters hunting for stories every day to embarrass the president and prove he wasn't up to the job. Those always guaranteed them attention on the cable news shows and made many White House reporters instant celebrities for "bravely" taking on Trump and "speaking truth to power." I hated that posturing, and to this day I don't like that phrase. In reality, many of them were far easier to work with in private than their public personas suggested and would even agree with many of the points I made to them in my office or on the phone. Indeed, some of the most outspoken or seemingly aggressive reporters were kind and reasonable people behind closed doors. I got the explanation "I have bosses, Stephanie" on more than one occasion. I wish I had been able to use that excuse with them.

As was customary in past White Houses, we had planned to

have "message" weeks for the administration, where we could focus the public's attention on a big policy goal and then work toward achieving said goal in Congress: Education Week, Jobs Week, and most notoriously Infrastructure Week. I say most notoriously because every time we told the press corps that we planned to focus on infrastructure, Trump would tweet something bonkers or some scandal or miniscandal would break or the president would undercut whatever message we were sending. "Infrastructure Week" became a running joke inside the White House, one we even shared with the press, a sign of the constant turmoil and chaos that would forever prevent us from doing a single damn thing we planned to do on a given day.

WE WERE ALSO A White House filled with inexperienced staff members, divided into different camps where many people leaked freely to reporters. That understandably upset the president, who was always urging some investigation to find the culprits. Peter Navarro led a few of them, most famously the search for the author of the "anonymous" op-ed in the *New York Times,* but Sean Spicer may have been the first to be tasked with "finding the leakers." It got to the point that in late February 2017, only a month after taking office, the entire communications and press teams were ordered to meet in Sean's office "in five minutes." No reason was stated, and the language in the email was less than friendly. When I walked in, half the team had already assembled and Sean was at his desk with two men I didn't recognize but who were soon introduced as lawyers from the White House Counsel's Office.

Once the entire team had crowded into his office, Sean spoke up. He said that there had been too many leaks from our morning comms meetings, which were held each morning in the press secretary's office to go over the news of the day and subsequent messaging expectations. The leak that had apparently set him off was that Michael Dubke had been hired as White House communications director, which the press got hold of ahead of the official announcement. I understood why he was pissed. If the White House communications team couldn't even keep our own internal personnel announcements under wraps, it was not a good look. Another problem was that Dubke was, in some ways, a divisive choice, yet another longtime DC establishment Republican operative whose hiring led to more alarms about "RNC people" taking over Trump World.

As I said, I did not disagree with Sean's (and the president's) frustration. Since we had taken office, there had been an unprecedented number of leaks. The Trump people blamed the RNC people because we had come from such a small, tight-knit group; in our minds it had to be the newbies. The RNC folks, on the other hand, largely blamed the Trump people because we weren't "sophisticated" enough to understand how to do things properly and, they assumed, talked too openly with reporters. The president had demanded that Sean find the leakers. As I said, little did we know that Sean would be the first of many, many White House employees who would be tasked with "finding the leakers" over the next four years—a role that would eventually be given to me. I myself knew of some leaks because reporters would tell me about them. Unfortunately, that put me into bad situations because I never wanted to burn the reporter, and so the cycle would continue.

After explaining the circumstances, Sean told us that we were to immediately turn over our phones—both personal and White House—to him or to one of the two lawyers present, so they could look through the text messages and phone logs and find specific reporters. "No one is required to hand their personal phone over," he said, before adding that it would certainly be noted if we refused.

To put it plainly, I was pissed. I had worked my ass off and had certainly never leaked, and to be treated in that manner by an RNC person (sorry, Sean!) rankled me further. I specifically waited in line to give my phones directly to Sean. I wanted to watch him look through my phone and see how often I texted with my family, my sick best friend, or my damn gynecologist. I wanted him to see how angry I was, never mind that common sense dictates that if you want to leak to some reporter you should maybe just meet him or her for coffee (I was not aware of encrypted apps at that point).

Deep down I knew that he likely had been given the directive, and I also knew that there was nothing on my phone—but I was still angry. A person doesn't give up everything and dedicate herself to an ideal so she can leak things later and be treated like shit by people who didn't put in nearly the work I had. I'm not sure if that makes sense or if it is fair, but it was how I felt at the time. That may have been the first in a very long string of instances when I thought somehow I was special—that I had shown my loyalty, put in my time, and deserved nothing but 100 percent trust. In that moment I couldn't contemplate that I'd ever be treated worse in terms of trust. Looking back, my ego and my fervent hope that we were all just working out the kinks was really just me ignoring my gut. I suppose that's why the whole "hindsight is 20/20" crap was invented; it still didn't mean I had to like it.

THE SEAN SPICER INVESTIGATION left a sour taste in my mouth—and maybe helped inform what happened next. On March 8, 2017, I was in my office in lower press when the first lady's operations director, Tim Tripepi, stopped by to ask if I would join him in the library in the Cross Hall, a historic hallway located in the bottom of the Executive Residence. Tim and I had worked together on both the Romney and Trump campaigns. In fact, he had been on my very first advance trip, a fundraiser in Los Angeles, and had taught me a lot. As we went way back, I figured he needed to speak with me privately on some topic. But as we approached the door to the library, he informed me that "some people from the FLOTUS office want to speak to you about potentially joining the East Wing."

Before I had a moment to process that information, we saw Sean Spicer walking down the hall, most likely to the medical unit. A necessary perk of being a part of senior staff meant having the ability to go to the medical unit to schedule doctor appointments at Walter Reed National Military Medical Center, get medical advice or prescriptions, or just request aspirin. Part of the unit's job was to keep the people closest to the first family healthy, because that was best for them and the country. Tim ushered me into the Vermeil Room, a room dedicated to the White House gold vermeil collection and portraits of previous first ladies, until the coast was clear. Since Sean was my boss and had given me a job that I still cherished, I didn't want to look disloyal, especially since I'd had no idea what the meeting was about in the first place. Once Sean was out of sight, Tim led me into the library, where I had a meeting that would change my life.

3

Rapunzel

Say yes to unexpected opportunities—even if it scares you.

—TORY BURCH

Like nearly every other person in the Trump administration and almost everyone in the outside world, Melania Trump was a total mystery to me. That was by design, I would later learn. She didn't want anyone to know her. She didn't care most of the time what people thought. There were maybe a half-dozen people who could say they had a true sense of Melania Trump, who had real insight into what she thought and did, and I was about to join that exclusive, jealously guarded list.

THE LIBRARY, WHICH WAS once used as the laundry room of the White House, is a square room on the ground floor of the White House, and the walls are lined with more than 2,700 volumes of American literature and history. The furniture is all dark wood with red upholstery. When I was called into my impromptu meeting with the first lady's office, I expected to see only Lindsay Reynolds, her chief of staff, and perhaps an assistant. Instead I was greeted by not just Lindsay but a group of impressive, intimidating-looking women. There were Rickie Niceta, the White House social secretary, Rachel Roy, a well-known clothing designer and model who was an adviser to Mrs. Trump, and the biggest surprise of all, the first lady of the United States herself.

My first thought on seeing Melania Trump in the flesh was that the woman was gorgeous. That observation may seem obvious, I know, but up close she is even more stunning. Because I know everyone is wondering, she even smelled incredible. She was impeccably dressed in a designer outfit, the details of which I can't precisely recall and couldn't do justice to even if I could. It was as if every article of clothing on her had been the outcome of a careful and deliberate decision-making process. Yet she wore them all comfortably and naturally.

What surprised me the most was how easy she was to talk to, how much she smiled and laughed, and how often she asked if I needed anything such as water or coffee. Her accent was heavy, especially for someone who had lived in the United States for more than twenty years, but lovely. I had little trouble understanding what she was saying, even if she occasionally used odd idioms. As she would throughout our time together, she would pepper sentences with questions such as "Do you understand what I'm saying?" just in case there was a doubt. I think it was a tic she had developed after so many years of being in the United States and still sounding so different.

The four women were sitting around a table with six brightly colored oval objects in front of them. Apparently I had walked into the final approval meeting for that year's White House Easter egg collection, and they were studying the various candidates and shades of colors as though it were the most important decision of their lives. That was the sort of stuff a first lady has to deal with all the time. To me, it was all so silly. Easter eggs? I mean, how many color options could there be—red, yellow, blue? But, as I would come to learn, anything involving aesthetics or decor was of utmost importance to the first lady. Image, fashion, design—that was the world she knew. Reflecting that, the East Wing team was known for their impeccable taste in clothing and their overall style and polish, and the women before me reflected that reality. Just like Mrs. Trump, Rachel Roy is so stunningly beautiful that it actually takes your breath away. Lindsay struck me with her sense of humor and how solicitous she was to Mrs. Trump, and I loved Rickie immediately, as she was warm, inviting, and attentive to everything I said. That was what made her such an exceptional social secretary.

In that beautiful room with those well-heeled women, I had never felt more unattractive in my life. I was wearing ill-fitting pants, worn-out flats, and a blue-and-white shawl draped over an otherwise grungy shirt because earlier in the day I'd been running the press pool around for opportunities with the president. Before me were women who wore Chanel and Ralph Lauren, whereas I typically grabbed sweaters out of bins at TJ Maxx (still do). Added to which I was a sweaty mess and my hair was in a sloppy knot on my head. I could have killed Tim for letting me walk into that room of beautiful women, who looked like they could walk onto magazine covers at any given moment. But not one of them seemed to register the sloppy state I was in or the shock on my face to be among them. They were probably used to it and graciously refused to make me feel awkward.

ONCE I WAS SETTLED and the introductions had taken place, they each asked me questions, most of them typical in terms of a job interview. Mrs. Trump immediately and quite innocently asked if I had just recently started at the White House. There was no reason she needed to know that I'd been there from the start, but it did seem a little strange that she hadn't been told basic information about a person she was going to potentially hire. I learned later that one of the reasons they wanted to speak with me was that I had helped Lindsay out on Air Force One when a story had started to circulate about Mrs. Trump leaving Akie Abe, the wife of Japan's prime minister, to tour DC all alone. Apparently it had been an honest mistake, and the FLOTUS team hadn't yet had a comms person to handle it. Lindsay had walked out of the office on the plane pale after having been yelled at by the first couple, and I had offered to get on the phone with a reporter whom I had gotten to know during the campaign. It was because I had been the only person on the plane who had offered to help that the FLOTUS team thought I could be a good fit.

Rachel asked some of the toughest questions. I wasn't aware of it at the time, but she was a successful businesswoman and philanthropist who had known Mrs. Trump from their time in New York City. I would also hear the rumors that some thought she was "Becky with the good hair" from the song "Sorry" by Beyoncé—but I never found out for sure. Looking back now, I understand where her questions came from, especially relating to how I would promote and protect the first lady's "brand." She was understandably protective of her friend and wary of letting in outsiders.

But that was also the problem. They were people totally unfamiliar with Washington, DC, and without much feel for communications, especially in politics. In that moment I realized that should I be offered and accept the job, there would be a big learning curve for the group. The world they came from had little to do with partisan battles and crisis communications. Since I had been thrown into this out of the blue and already felt self-conscious, hoping they weren't studying me as if I were some sad clown, I just went for it.

"With all due respect, I think your potential is being wasted," I told the first lady. "I'm personally proud that we have a first lady who is so independent and a badass." Yes, I threw that term in there.

Mrs. Trump looked expressionless, although maybe I detected a slight smile on her face. She didn't respond directly to that, and I couldn't tell if she was amused or impressed—maybe a little of both.

I offered a few simple examples of policy that could be easily implemented, truly impactful, and quick ways to garner positive publicity in a fairly short amount of time. One would be to find state legislation that was aligned with her policy preferences, usually involving children, and publicly support it. If the legislation passed, she could make a trip to the state for a ceremony to celebrate the new law and then encourage replication of the law in other states. That would be a tangible success to point to, which would show the American people that she was making a real impact on behalf of children.

Now I was the one who felt as though I were speaking a foreign language. As I kept talking about how the Washington press worked, ways to respond to stories or gain attention, and how to put forward a policy agenda, the others looked confused and frankly unimpressed. Never did any of the women ask how they could help the president's agenda. Nor did they ask me for a strategy to assist with doing so. Nor did they even talk through a strategy for the first lady, but maybe that was because she didn't have any initiatives that had been announced yet. I thought I might be blowing the entire thing, but I hadn't been given time to prepare for an interview, so it felt good to just be direct.

Thankfully, Mrs. Trump had a way of making you feel at ease, as if you were immediately "in the club." She seemed genuinely interested and asked many questions about my family, my background, my goals, and my thoughts. I was surprised, to be honest, because the Melania Trump most people see is quite formal, not often smiling, and mostly very quiet. That was not the person I met that day in the Library, and as the minutes wore on, I found myself hoping that I would be offered the job—to get to know her, learn from her, and also do a good job on her behalf.

I'll admit that I was also curious about the relationship between the president and first lady. As with everyone else, there were rumors—it was a marriage of convenience, they didn't get along, and so on—but nobody really knew. I wondered why that beautiful, elegant woman was attached to an older guy who wasn't necessarily the best looking, though he had been okay back in the day, I suppose. Was it really just the money? She certainly didn't offer any clues during our first meeting. That mystery was something left for me to uncover over time.

AFTER THE MEETING WAS over, Tim told me, "You did great." I wasn't so sure. I didn't really know what to think. But within a day the first lady called me personally. She began the call with what I would learn was her signature "How are you?" that began every conversation. It wasn't always clear that she was interested in the answer. Then she dived right in. "I would like to offer you the job as my communications," meaning her spokesperson.

I told her I was honored, and she outlined the duties. "You would send out the press releases and talk to the reporters and get to know them," she said, "and help me with the social media and my initiatives. Do you do TV?"

She then quickly turned to the plan for announcing me as her spokeswoman, saying that I needed to start right away. Out of deference to Sean Spicer, I told her that I wanted to give him a month's notice. I knew how hard it was to hire someone for the

White House staff. They needed to be vetted, get a background check, and move through the ever-slow bureaucracy of the federal government.

"You will tell them two weeks—and nothing more, they do not need it," Mrs. Trump replied sharply in a tone that suggested that answering anything other than "Yes, ma'am" would not have been acceptable. That was the way it would be. The edge in her voice caught me by surprise. But, yes, I started two weeks later.

ONE PERSON WHO WAS not present at my interview with the first lady was a woman named Stephanie Winston Wolkoff, and Mrs. Trump told me that I should talk to her as soon as possible, that she was her adviser and "getting the initiatives together." What initiatives? you ask. I don't think anyone was quite sure. I only knew that a woman named Stephanie had been helping Mrs. Trump set up the East Wing and would be a temporary senior adviser as we worked to get the office up and running. She lived in New York City and had recently been hospitalized for a back surgery. So I emailed her to say hello and that I looked forward to working with her—the usual pleasantries, at least on my part, and ones that the first lady had asked me to convey.

It was years later, after Wolkoff had written her own book, that I learned my hiring had upset her and that my reaching out had put "chills down her already injured spine." Despite having never met me, she somehow purported to know that I was dishonest and also pointed out that I was very connected with the press. That was meant to suggest to the first lady that I was some sort of untrustworthy leaker, but having relationships and working with the press was actually the whole point of my job. She also took a delightful jab at me when she wrote that she had "wondered if the First Lady knew of my past legal issues." Classy.

I am going to struggle to limit my discussion of her in this book, but it will be hard. And you, the reader, need to know about her because it says something about some of the people who surrounded the Trumps from the very beginning. In short, many of them were wealthy grifters from New York, people who claimed to know what they were doing and made huge promises they couldn't deliver. That was in part because the Trumps had no political experience and in part because they had egos that a certain subset of the wealthy and elite were able to manipulate. Among the entire cast of characters in the Trump administration—and it did sometimes feel like that, a cast on a TV show—Stephanie Winston Wolkoff was to me, one of the most puzzling. I met with her in person maybe three times—maybe. One of those times she had called a meeting to discuss Mrs. Trump's "initiatives," which hadn't yet been developed. Stephanie was big on "emotional intelligence," which is indeed very important when it comes to children's well-being. The problem was that she was the one with a passion for it, not Mrs. Trump, who continued to tell us that online issues and drug abuse were her focus. Wolkoff was relentless, though, and that day she had put together some presentation on the topic, using a bunch of ten-dollar words that in my opinion she didn't understand and that would never translate to the general public. And although I'm all for different points of view and ideas, I'm not for someone refusing to hear anything but her own voice.

What's more, it seemed that Stephanie saw conspiracies everywhere, and perhaps her biggest legacy was inserting into the first lady's mind that she was somehow being mistreated or undermined by her husband's staff. From my minimal interactions with her, it struck me that she set a tone from the beginning that would pit the East Wing against the West Wing, a conflict that never fully resolved itself over the next four years.

The centerpiece of most of Stephanie's conspiracies was the first daughter. She was apparently convinced that I was some sort of spy for Ivanka, something Mrs. Trump and I later laughed about many times. But her disdain for Ivanka was not hers alone; it permeated the East Wing. That was my first glimpse of the potential tensions between the president's wife and his daughter, although the complainer in chief at the time was Stephanie.

Wolkoff did have a big role in the beginning, and I could see that she had the frenetic energy needed to get the Office of the First Lady set up. And I'll give her this: she was right when she later wrote in her own book that everything the Trumps do is "transactional." I watched for more than four years the way that in any moment, the Trump family was able to convince the world that any problem was due to "everyone else" and they were never to blame. Omarosa Manigault, General John Kelly, Rex Tillerson, Anthony Scaramucci, Stephanie Wolkoff, Cliff Sims, General Mark Milley, John Bolton, Bill Barr, Mitch McConnell, Vice President Mike Pence, and on and on and on—it was always their fault, they were disloyal or seeking attention, profiting off the Trump name, incompetent, pathetic, or a loser. And I generally always believed the Trumps—that everyone else was in the wrong—even when my gut told me not to.

MY NEW POSITION AS the first lady's communications director was announced on March 27, 2017, and the support I received from the West Wing was overwhelming. Sean Spicer and Sarah Sanders were both amazingly gracious, as were my campaign friends and all my comrades in the advance world. A big highlight was when I entered my office in lower press the next day to find a newspaper on my seat. It was a copy of the *Washington Post* with the story of my new job on which the president had written STEPHANIE—YOU

WILL DO GREAT!! in his signature Sharpie marker in all caps. The best part was the arrow he drew pointing at the picture of me— you know, in case I didn't know it was me.

My first days in the Office of the First Lady were confusing, fun, and, frankly, relaxing. It was a close-knit group who laughed together a lot—and the ten-hour days I had been working in the West Wing disappeared. The East Wing personnel were all fiercely protective of Mrs. Trump and focused on helping her succeed. The team was small, consisting of the chief of staff, who also served as her personal aide, the operations and advance team, a couple of administrative roles, and the social office, which included the social secretary and her staff of two. There was not a dedicated policy role when I began; that was left to Stephanie Wolkoff and Pamela Gross, another New York friend of Mrs. Trump who had previously worked as a producer for CNN.

AS THE DAYS AND weeks moved along, it became clear to me that Mrs. Trump did not really understand my role as a spokesperson. To be fair, this was true of all the Trumps, from the president on down. As I've mentioned, they were from the world of celebrity, not politics. They saw the role of a press person as that of a personal publicist, not unlike those they had used in the past to plant short favorable stories in places such as Page Six and *People*, as other celebrities do routinely. Do you think that all the great stories about what a nice guy Tom Hanks is show up by accident? Before he entered politics, Trump quite famously once served as his own publicist, creating a persona named John Barron, who would call up celebrity or financial reporters at magazines or newspapers, sometimes embellish or make up things, and get them to print pretty much whatever he wanted. When you brought reporters to a new Trump property and gave them elaborate fruit and cheese plates and a free day of golf, you got puff pieces more often than not, especially when the stakes were low. The host of *Celebrity Apprentice* was held to a different standard by the press than a president or a first lady is. It's a totally different dynamic, and it took a long time—a very long time—for the Trumps to understand that. Maybe they never did.

Similarly, the first lady's experience with the press was largely shaped by her career in modeling. In that world, photos were far more important than what was said in print or on television. Master publicists could kill unflattering pieces, and she'd had the ability to approve whatever statements she offered to a reporter. Interviews were scripted and rare. It didn't take me long to realize that the entire Trump family had generally used marketing and advertising to sell their brands, meaning that they had paid publications and networks to run commercials or ads that said nice things about them or their products. Not to belabor the point, but that simply is not how it works in the White House.

So from the very start Mrs. Trump struggled to grasp why I could not tell reporters what to say or to write. I tried to explain that that would be propaganda, but my explanations only frustrated her. I had to go through the basics with her—and the entire East Wing staff—about what it meant to be off the record, on background, and on the record and what a press pool can and cannot do. It was truly Communications 101 stuff—and after three years, she and most of the staff still didn't understand it. Or want to accept it.

She had no sense of time pressure. The 24/7 political news cycle was completely foreign to her. She didn't understand that sweating over the perfect press release actually undermined her purposes because reporters wouldn't wait—the news cycle wouldn't wait—until she'd figured out the perfect statement. By the time she was ready to respond to something, the story was already done, the damage was done, and the press had moved on. I could see that she didn't know how to value which political reporters deserved attention, which news stories were problems, and which didn't really matter. Although neither would enjoy the comparison, she was similar to Ivanka in that way. Like her husband and all of his kids, Mrs. Trump scrutinized her press clippings like an expert architect focusing on blueprints. No detail was overlooked, nothing missed her eye. She had Google alerts set up for herself and saw everything. There were times she'd be angry over a story and tell me that "it was everywhere" when I had not even seen it yet. I also learned quickly that I was expected to send *every* press inquiry to her. In the beginning I wrote out statements and sent them to her for edits or approval, but I soon learned that it was better to see if she even wanted to respond, because 75 percent of the time she would just say, "Don't replay," which, in her occasional garbling of the English language, actually meant "Don't reply."

The effect of that behavior, or misunderstanding, of how the political press worked was that it allowed others to shape her image in decidedly negative ways. After massive flooding took place in Texas in 2017, Mrs. Trump decided to travel there to visit with victims and offer support, a normal action for a first lady to take. But she objected to my telling one magazine on the record that the first lady planned to "continue supporting all those who were affected by natural disasters." The magazine was doing a piece on her travels there, and I thought that was a pretty benign answer. But no.

Instead, Mrs. Trump asked to meet with me in the Map Room. This is one of the smaller ceremonial rooms in the White House, on the ground floor, with low ceilings and filled with uncomfortable colonial-style furniture. The room still contains a couple of maps that were there during World War II, when President Franklin D. Roosevelt used it as a place to strategize with his generals. That day we were having a strategy session of a slightly different kind.

The first lady had seen my quote and didn't like it. "I feel like

I'm losing my own voice," she told me. She didn't want anyone speaking for her, even though that was my job. The problem was that she never said anything on her own. As a result, her well-intentioned visit to Texas was dominated by a story of her wearing stilettos from the White House onto Air Force One. That, of course, played right into the emerging narrative that she was an out-of-touch, frivolous elitist who didn't care about people. There was some truth to that image—most rich people are out of touch, in my experience—but it wasn't true that she didn't care about people who suffered.

All of that was an odd and difficult adjustment for me, because as a communications person, most certainly in politics, it was second nature to immediately answer reporters on behalf of my principal, whether on the record, on background, or to give off-the-record context. It is PR 101. But again, not with Mrs. Trump. And again, her own past experience with the press did not provide her much guidance. A model tends to speak, or more often not speak, for herself.

I later learned from her own book that Stephanie Wolkoff had convinced the first lady that my efforts to speak to the press on her behalf were attempts to build my own image and public persona—that I was, in effect, trying to become a star on her dime. After that incident, I never responded to a reporter in *any* capacity without her permission—and you better believe that that would place me in some tough spots professionally, especially with the president.

The irony was that even though my ability to speak out as a spokesperson was constantly constrained, there were sometimes situations in which I was deployed to speak when silence might have been better. In fact, I gained a reputation as a combative press spokesperson who put out a bunch of tough statements, generally over silly or unimportant things that Mrs. Trump tasked me with responding to. One infamous example involved the labored, agonizing response of the third Mrs. Trump to the first Mrs. Trump. That was my first crisis on the job—if you can call it a crisis—and an invaluable lesson on how things worked in Melania World.

IN THE FALL OF 2017, Ivana Trump, the mother of Don Jr., Ivanka, and Eric, published a memoir entitled *Raising Trump*. I really hadn't heard Ivana's name up to that point or any mention of the president's second wife, Marla Maples, for that matter. But the first Mrs. Trump was silent no longer. In her new book and publicity tour, she said a whole bunch of kooky things, such as that she had encouraged her kids to have playdates with Michael Jackson, of all people.

It became my problem, though, when Ivana sat down with ABC and bragged about having a "direct number" to call the White House. But she didn't like to use it, she claimed, "because Melania is there." She elaborated, "I don't want to cause any kind of jealousy or something like that, because I'm basically first Trump wife. Okay? I'm first lady."

The third Mrs. Trump watched all of that play out on television and social media—remember, she never missed anything—and fumed. She seemed to think that Ivana Trump was a ridiculous person who was "only out for attention" and grossly exaggerated how much time she spent talking to the president. It was one thing for her to be out there making up titles for herself—"first Trump wife," whatever the hell that meant. But it was clear that Melania really hated the "I'm first lady" remark, for which I didn't blame her.

There were a number of reporters anxious to see if Mrs. Trump number three would go after Mrs. Trump number one. The first lady, of course, saw all of the many Ivana inquiries we received and directed me to ignore them: "Don't replay."

But as more and more media attention descended on Ivana, Mrs. Trump began to have second thoughts. "Should we say something?" she asked me one day.

I replied, as I would do with every Trump-related book that would come out in the next few years, that staying silent was the best course of action. "You will only help her sell books," I said. She agreed, or at least she seemed to.

At some point, she could no longer hold her tongue, and she asked me to draft a statement in my name that she would edit. The response—which one publication called "tart" and the entire world predictably seized on—said the following: "Mrs. Trump has made the White House a home for her son and the President. She loves living in Washington, DC, and is honored by her role as First Lady of the United States. She plans to use her title and role to help children, not sell books. There is clearly no substance to this statement from an ex. Unfortunately only attention-seeking and self-serving noise."

The irony was that the first Mrs. Trump had spent most of her time going after the second Mrs. Trump, Marla, who had apparently ruined her marriage. Fair enough. She had largely left the third Mrs. Trump alone, except to commiserate with her. "Frankly, I wouldn't want to be in Melania's Louboutins right now," Ivana once wrote. "No, I don't want anything to do with Washington. What do people do for fun at the White House? Throw bowling balls in the basement with security guards watching your every move? Forget it." At the time I had no idea just how closely my new boss shared that sentiment.

A UNIQUE ATTRIBUTE OF the first lady's personal offices in the East Wing was that they did not include the first lady. When I first was given a tour of the suite of rooms, I was shown a large, beautiful office designated for Melania Trump. This is the office in which Michelle Obama, Laura Bush, and any number of their predecessors set up their base of operations. The room had a desk, two chairs, a couch, and a pristine white carpet. (Pristine because it rarely saw foot traffic.) Lindsay and the others told me that the first lady rarely made an appearance there. In fact, I could count only a handful of times over the years that Mrs. Trump was actually in her office. She preferred to run things via text or phone call, which initially made it hard to set an agenda and form a close working relationship. Mrs. Trump was working from home long before the country was. When warranted, we would have in-person meetings, but those generally took place in the Map Room across from the elevators to the residence. There we would plan out schedules, respond to pressing queries, and discuss goals. Other than that, the first lady kept to her rooms in the residence. That became a running joke among those who knew her. The Secret Service unofficially dubbed her "Rapunzel" because she remained in her tower, never descending. In fact, some agents tried to get assigned to her detail because they knew the first lady's limited movements and travel meant that they could spend more time at home with their families.

The first lady told us all the time that she was "extremely busy." Busy with her son or her parents, who were almost always with them. Busy with renovations to the White House or Camp David. Busy planning for the next big White House event, such as the Easter Egg Roll, a Christmas reception, a state dinner, or an event in the East Room. I would find out much later that she was also busy with her photo albums. As a result, she rarely agreed to attend public events that required travel outside the White House. In some of our meetings, we considered it a success if we got her to agree to one activity per week, which was always our goal. We would walk away thrilled after she signed off on a slew of solo events, only to be crestfallen when she canceled the events she'd just agreed to one by one. Eventually we learned to stack events, which meant that when "Rapunzel" was certain to descend, dressed and with her hair and makeup ready, we would pack her day with everything we wanted her to do that week or month.

One example of how extreme her habit of staying hidden was came a few weeks after the tragic shooting at a baseball practice for the annual Congressional Baseball Game in June 2017. One of the people who was shot was Steve Scalise, the House majority whip and a key ally of the president. After he recovered, Scalise and his family came to the White House for an impromptu visit. They were gathered in the Blue Room, and we asked Mrs. Trump if she would like to come down from the residence to greet them and take a picture. Her response was "No, I already said hello," referring to the fact that she and the president had visited Steve while he had been in the hospital weeks earlier. That would become a running joke in the East Wing. Whenever the first lady said no to something, we would say to each other, "Well, she already said hello!" Say what you will about Ivanka and Jared— and I'll have a lot to say as this book goes on—at least they showed up in their offices most every day to work.

When Mrs. Trump did undertake a public activity, we invariably found some reason to regret it anyway. When she assumed the role of first lady, there was speculation in the press that Mrs. Trump planned to do away with the vegetable garden that Michelle Obama had planted. I have no idea where the rumor originated, but a number of reporters seemed convinced that the first lady planned to gut the place and perhaps erect a gold statue or something. That was never true. To demonstrate her commitment to the garden, we arranged an appearance with a local children's group, when Mrs. Trump, who truly loved kids, could show them the White House grounds and talk about the importance of gardening and eating healthy. Mrs. Trump was disappointed, as was I, that the main takeaway of the news coverage was her pristine white shoes and the $5,000 flannel shirt she was wearing. What, the first lady wondered, was the point if all people cared about were her clothes? She wasn't wrong. But I believe that only further strengthened her determination to avoid making public appearances whenever she could.

OVER THAT FIRST YEAR working for her, I also grew aware of her vulnerabilities. Every once in a while she would let her guard down with me, revealing her anxieties and preoccupations. Although I thought her accent was exotic and even beautiful, she seemed acutely self-conscious about it. She knew she didn't sound like a "regular American," and she saw people mocking her way of speaking on late-night shows and *Saturday Night Live*. Though she never quite said it, it clearly bothered her. I think because she was unusually attractive, which people privately envied and resented, and also because she made no effort to engage on a continual basis with the press or the public, people thought it was fair to belittle her accent in ways they wouldn't have done for other women in the public eye. I hate to say it, but can you imagine if the tables were turned and a Republican dared to make fun of someone's foreign accent? The hypocrisy was palpable. Although she could make fun of her own speaking style in private, she wouldn't allow herself to be self-deprecating in public. Because she was uncertain about idioms and expressions, she rarely wanted to write anything on her own. I used to write thank-you notes, tweets, responses, speeches, remarks, even condolence letters for her—not because she didn't have the time or because she didn't care, though sometimes that was the case, but because she wasn't confident of her grammar. On the rare occasions when she did write anything on her own, she usually gave it to me before sending it to be sure it was grammatically correct.

In July 2017, I sat with her in her cabin on Air Force One on the way back from Poland, where she had just delivered remarks and introduced her husband in Warsaw's Krasiński Square. On the plane, she asked me to sit with her and go through every line of the speech to identify what words she had mispronounced. It turned into an hour-long English lesson of sorts on spelling, phonetics, grammar, and punctuation. It was one of the rare moments when she not only admitted what she perceived to be an imperfection but I felt she trusted me enough to let me see that. In fact, it's hard to write this right now. But it is also something I think people should know, because I believe they would have related to her a bit better if they had. I know I did.

THOUGH "RAPUNZEL" WAS THE Secret Service's unofficial name for the first lady, her actual code name—"Muse"—was also fitting for a different reason. A muse is a source of inspiration, but at least as I think of it, more of an object or an idea than an actual person. In that respect, Melania Trump was whatever people wanted her to be. She was not very talkative or reflective. She rarely shared her inner feelings or thoughts. I learned to decode her silences, her eye rolls, her sighs, her shrugs, the movements of her head. It got to where I could walk into a room and just by looking at her decipher if she was happy or irritated. She preferred to react to what we said than for us to react to her. Perhaps because she was uncertain of her English, she stuck to five or six standard phrases and sentences that she repeated all the time. I was as close to her as anyone on her staff, or so I believed, but I was never certain that I had figured her out.

East Wing events during the first year of the administration were scattered. The plans for the first Easter Egg Roll were under way, she spoke at the State Department's annual International Women of Courage Awards, we visited a local charter school with Education Secretary Betsy DeVos and Queen Rania of Jordan. I noticed immediately that the first lady's interactions when the press were around versus when they were not were vastly different. With the press, I could see her working to ensure that her face was held a certain way, that she stood or sat a certain way—she was a model through and through, which made for great photographs but also gave off a chilly vibe to the people she was meeting with. But after the cameras left, she relaxed. She would laugh and ask questions, kneel down to speak to children, and generally behave the way I knew her to really be—it was the Melania Trump I loved to see and wished the country could have seen more of.

OUR FIRST BIG SOCIAL event was the annual Easter Egg Roll at the White House, a tradition that dates back to 1878, where the president and first lady welcome children to the White House lawn to roll eggs, play games, and hang out with their families. It was also the first large social event of the administration. The *New York Times* headline suggested that eyes were already starting to roll along with the eggs: "The Latest Test for the White House? Pulling Off Its Easter Egg Roll." Other outlets quoted people saying that it was the "single most high-profile event" at the White House each year and the administration and first lady would most certainly be judged by it. So, no pressure. (Now I could better understand why the colors of the damn Easter eggs were so important.) That was the "mean girl" part of Washington, DC: pundits and society types were just waiting to see if Melania Trump and her skeleton crew could handle something like this, and it felt as though they were not-so-secretly hoping for a flop.

Mrs. Trump knew that. She felt the pressure. Everyone around her did. But if there is one thing she was good at, one thing she had experience in, it was making something look spectacular. She had hosted many functions as Mrs. Donald Trump, after all. When planning the Easter Egg Roll, she made some requests that ended up making the event more successful than anticipated, one of them being to keep the crowd size smaller than was customary. Unlike her husband, crowd size was of no interest to Mrs. Trump, instead she wanted to focus on the quality of the experience for the children and their families. In her mind, standing in line forever was not fun for anyone involved—and it turned out she was right. Paying attention to every single detail, she even scrutinized the Easter bunny. Just as the guy in the costume was about to go onto the balcony to wave to the kids, a frown crossed the first lady's face. "I don't like what he is wearing," she said, eyeing the plaid vest the bunny had on. She felt it was tacky and distracting. Then and there, only minutes before he was to hop out onto the White House lawn, Melania Trump made the Easter bunny strip.

THAT DAY ALSO INVOLVED the other members of the Trump family, an often fraught dynamic in the coming years. Don Jr., Eric, and their kids all made sure to be front and center. They cut the line ahead of the other children in attendance and made sure their own kids and not the kids of the general public were in the photos with the president and first lady. Some of the children and their parents had stood in long lines for an hour or longer to get a spot with the president, and they were relegated to the background. I would go on to develop positive relationships with most of the Trump children, but that was a generally obnoxious and entitled display that did not appear to surprise the first lady, Lindsay, or other East Wing aides in the slightest.

That was also the first time I picked up on the potential for tension between the first lady and the elder Trump kids, whose behavior at the White House she sometimes deemed inappropriate. Appropriate conduct, or her definition of it, was a very big thing. When Eric and Lara wanted to hold their son's christening at the White House, she expressed concerns about the optics. She had good instincts. But the christening went ahead anyway.

AS THE YEARS WENT on, the Easter Egg Roll became one of my favorite holiday events. The planning that went into it was intense; I will never see another one again and not appreciate whatever administration worked on it. Each year had its own challenges. One year it rained. One year someone thought it would be a good idea to serve hard-boiled eggs on a stick to the children; they were disgusting.

On another occasion, I happened to be out on the lawn when I heard in my earpiece that a man had suffered a heart attack near a station that the president and first lady were about to visit. The Trumps were not exactly comfortable with anything to do with illness or death. The man, who was apparently shirtless and being worked on with paddles, was foaming at the mouth—a sad scene. So I took it upon myself to help the president and first lady avoid walking right over to it. Just after the president had blown the whistle for the Easter Egg Roll, I hurried over to the first couple, in front of a huge bank of cameras and the press corps. I got their attention and pointed my finger in the opposite direction of the man being resuscitated. The Trumps instinctively turned their heads to where I was pointing. I smiled the whole time as I said, "Mr. President, we are going to head over this way. There is a man down behind you. He had a heart attack and they are working to save him, so we're going to take a different route."

Totally confused, Trump asked, "Is that real? Is that real?"

Of course it was real, I thought. I didn't just make that up for no reason.

He didn't ask if the guy was all right, though. As far as I know, that never crossed his mind.

4

Shangri-La

Let's just go and not come back for a while.

—UNKNOWN

As the first year progressed, our schedules and travels assumed a certain rhythm. Many of my weekends were spent at Mar-a-Lago; nearly every time she or he went, I went also. I was there most weekends during the "in" seasons. And I was there on the holidays: Easter and Thanksgiving and Christmas and my favorite, New Year's Eve.

Mar-a-Lago, or the "winter White House" as the president liked to call it, is a beautiful and unique place. The main house is in the middle of the property and overlooks the ocean on one side and the Intracoastal Waterway on the other. Though the rooms in the main house are quite dated, they are still distinctive and alluring. The property also has cabanas by the beach and pool, as well as cottages near the main building that have more than one bedroom and kitchenettes. One of the cottages was always reserved for the chief of staff because it had been wired for secure communications.

The architecture of the place is Spanish inspired with a tiled roof, built in the mid-1920s. Two notable architects, Marion Sims

Wyeth and Joseph Urban, were commissioned to build it by Marjorie Merriweather Post of the famous cereal family. It's the second largest mansion in the state of Florida and one of the largest in the country.

Mar-a-Lago became like a second home to me, as I had been going there since the early days of the campaign. The staff are incredible and greeted me with hugs and whenever possible my favorite cabana by the ocean. I grew close to two of the housekeepers there, as well as a few others who worked in the "front of the house." We all worked together to ensure that we were on the same page with schedules and guests for both the president and first lady, right down to their dinner courses every evening. Tuesday through Saturday nights, the open-air patio of the main house would be packed with members, many in outlandish but very expensive outfits. On the weekends, there was a singer who performed when the first family was dining; she knew all the songs the president liked the best—many of them show tunes: "Memory" from *Cats* and "Don't Cry for Me, Argentina" from *Evita*.

Mrs. Trump was usually there with their son and her parents, Viktor and Amalija Knavs. The Knavses were wonderful people and totally uninterested in the political and media crap that the Trump side of the family was obsessed with. As Mrs. Trump's father didn't speak English that well, he would walk up to everyone and say, "Hello, I am Viktor. Father of Muse" (the first lady's Secret Service name). Mrs. Trump's mother was a striking woman, and I could see where Melania got her looks. Mrs. Trump was very close to her parents, as was her son. The Knavses were their refuge. At Mar-a-Lago they had their own suite, and sometimes Mrs. Knavs would cook for them. Viktor had a car and was often out washing it with a bucket and sponge. He didn't need to do that, of course, but that was his way and it was charming.

*   *   *

SOMEONE WHO WAS ALMOST as frequent a guest at Trump properties as I was was Senator Lindsey Graham. Okay, I'm exaggerating, but not by much. The difference, of course, is that I was there to work. Of all of the various characters and hangers-on in Trump World, Lindsey was one of the weirder ones, and that's saying something. He seemed almost schizophrenic. Some days he would be one of Trump's most vigorous defenders; other days he was a harsh critic. People around the president would tell him that he couldn't trust Graham, but Trump seemed to like him for whatever reason and I often wondered if he sought Lindsey's approval. And Lindsey Graham? Well, it struck me that he was using Trump to mop up the freebies like there was no tomorrow (seems that he still is). He would show up at Mar-a-Lago or Bedminster to play free rounds of golf, stuff his face with free food, and hang out with Trump and his celebrity pals. On one occasion, I came across him at Bedminster after he'd kicked out a White House staff member so he could take her room. Senator Freeloader was sitting at a table by the pool, a big grin on his face, lapping up the goodies he was getting like some potentate. He said to me, with a creepy little smile, "Isn't this great? Man, this is the life." I remember thinking "Yes, it is, but it's not your life—it's the president's life." There was something so gross and tacky about his behavior during that trip that the image never left me.

IN APRIL 2017, I helped advance the Chinese visit to Mar-a-Lago. It was not a formal state visit, but it would be the first large foreign visit since Trump had taken office, and of course the president wanted to host it at his famous property in West Palm Beach. That posed many problems, one of which was that taking the Chinese out of the security of the White House grounds could be an issue. The State Department felt it was against protocol, and the resulting logistical challenges provided me with my first fascinating foray into security operations.

Almost as soon as the visit was officially announced, the hotels in West Palm Beach were suddenly completely booked for a "Chinese wedding" that was taking place in town, which meant that half the Chinese government was going to be in town trying to gather intelligence. Rooms at Mar-a-Lago had to be swept by the Secret Service for devices constantly, and more than once I was told to keep an eye on the press pools because there would be Chinese cameras pointing not at the official events but at all of us to try to gather intelligence.

One funny suggestion during the visit came from Joe Hagin, who was the deputy chief of staff for operations at the time. Joe proposed a shot of the two presidents, Trump and Xi Jinping, walking barefoot on the beach together. That idea was nixed pretty quickly—Donald Trump didn't show his bare feet to just anybody—and replaced with a lovely shot of the two couples walking across one of the lawns at the private club, with the ocean in the background.

After the advance portion of the visit was concluded, our Chinese counterparts took us to lunch. I found it odd that they insisted that we sit in specific seats and downright creepy when I found that my tablemates knew a whole lot of personal information about me. I am all for doing a quick Google search before you sit next to someone in a professional or social situation, but that was much more in depth. The women who sat at my table knew the names of my family, what jobs I had held in Arizona, and which street I'd lived on.

The official visit itself went well. Mrs. Trump took Mrs. Xi to a school that focused on music, which we were told was something dear to Mrs. Xi's heart. A beautiful dinner was held at Mar-a-Lago, and I got to watch President Trump brag about the "most beautiful chocolate cake you've ever seen"—in reality, though it was a good, large slice, it was pretty average—while talking about a missile strike on Syria he'd launched the week before. That, too, was deliberate. There was a geopolitical angle to the whole affair and in my opinion the president wanted to show off, and perhaps even intimidate, the Chinese.

CAMP DAVID PROVIDES A different, more rustic kind of luxury than Mar-a-Lago, but it is a truly special place. It is an hour drive from the White House or a twenty-minute helicopter ride. It is the spot where Dwight D. Eisenhower, who named the place after his grandson David, hosted the Soviet leader, Nikita Khrushchev, for critical Cold War talks. Two major Middle East peace deals were negotiated there—by Jimmy Carter in 1978 and Bill Clinton in 2000. Richard Nixon had the pool put in. Ronald Reagan invited British prime minister Margaret Thatcher to stay there. George W. Bush hosted British prime minister Tony Blair there four times. It is a place where presidents and their guests can be themselves—or at least relax a little bit—away from the Washington hustle.

I was lucky enough to go to Camp David several times, and staying in a place with such a rich history was humbling. The members of the military who live and work at Camp David are entirely devoted to the first family and senior staff. It is a place to get work done in silence, rest, and escape the media for a few days. It is the one place that the president goes without his press pool. Instead, one still photographer is on the grounds when the president arrives. He or she is not allowed to take pictures; it is just to record that the president arrived safely.

The first time I went to Camp David was in September 2018, along with Mrs. Trump's chief of staff. We wanted to get up there and take a look at all of the cabins and various gathering spaces in case she ever decided to go. In addition to scoping out the first family's cabin, which was extremely "rustic" but spacious and had its own pool and a driving range, we looked at the guest cabins, all of which had been newly renovated and redecorated. The guest cabins are beautiful and well appointed, right down to the coffee makers and the toiletries in the bathroom. Each closet holds various coats and sweatshirts, all of which visitors are allowed to wear while there, with the option to purchase them before they leave. For me, the best part of the guest cabins is the guest logs dating back to the 1960s showing who stayed in the cabin before you and the purpose of the visit. I saw names on the log such as Vladimir Putin, King Abdullah and Queen Rania of Jordan, and Beyoncé Carter and her daughter, Blue Ivy. Once I had visited a few times, I even saw my own name in the books; that was when I really knew I was part of history.

Mrs. Trump, however, didn't seem to be a fan. Even though she appreciated Camp David's history, the words "rustic" and "outdoorsy" were not words that described her lifestyle. Like ever. Her idea of being in touch with nature was sitting, occasionally, on her private balcony at Mar-a-Lago or Bedminster.

Camp David seemed to do all it could to reciprocate those feelings. On her first visit, the first lady went to turn on the shower, only to find it ice cold. The president had used up all the hot water. That sort of thing didn't happen at a Trump property. The camp promptly installed larger hot water heaters on the property so it wouldn't happen again. But the damage was done; from then on, the first lady would merely tolerate Camp David. She would never love it.

The president had his own issues with the water at Camp David. Every time we were there, he complained about the "soft" water and the water pressure, both of which he claimed were bad for his hair and made it so he could never comb it right afterward. Considering what an ordeal he undertook every day to get his hair to look the way it did, that seemed like a legit complaint. Donald Trump's hair, when not perfectly coiffed, was a sight to behold. There is no way to describe exactly how he styles that magnificent and very wild mane of his, but it involves a comb, a hair dryer, and a shit ton of hair spray. His hair is much longer than I had imagined, like multiple inches from end to end. He cuts it himself with a pair of huge scissors that could probably cut a ribbon at an opening of one of his properties. As I said, he had a thing about water and water pressure—not just showers but toilets, too. It was one of his constant go-tos. So at the president's direction, Dan Walsh, then a deputy chief of staff for operations, had water samples taken from Mar-a-Lago, Bedminster, and Trump Tower so that staff could determine the ideal water softness/hardness and replicate it at Camp David and ideally also at the White House—and then the Trump Organization could do the same across all of his clubs. I have no idea what they eventually landed on, but that's how big of a deal the water business was.

WHENEVER THE PRESIDENT WAS at Camp David, the schedule was pretty much the same: get some work done during the day, then have dinner together in the evening, followed by a movie. The movie theater at Camp David was a large room filled with huge, comfortable reclining armchairs. Popcorn was always available, and blankets were handed out. It was almost like you were sitting at home watching a movie with your great big family. The president always sat in the front row with his Diet Coke and popcorn, next to the camp commander and his family. I don't remember Mrs. Trump ever attending. The first movie I watched at Camp David with the president was *Sunset Boulevard*, which I had never seen and was dreading because I am not a fan of old movies. The boss was excited all through dinner beforehand and couldn't believe I had never seen it. I must admit, I loved it and was shocked at all of the similarities between President Trump and Norma Desmond, the lead character in the movie, who was a former silent-film star obsessed with her looks and with making a triumphant return to the screen. Here was a woman who was convinced that everyone loved her and lived in a fantasy world of her own making. I'm sure that Trump had no clue—like none—how similar to him she was. Still, the president was thrilled that I enjoyed the movie so much, and it would be a frequent topic of conversation between us over the next few years.

One of the other movies I watched with POTUS in the theater was *Rocketman*, a biopic about Elton John. What I remember from that night, however, was that the camp commander's daughter, who was maybe twelve, was sitting next to POTUS. There was a particularly racy sex scene depicting an orgy in a nightclub, and I felt uncomfortable. It was like sitting in the room with your parents when a sex scene comes on. After the movie was over, the president asked the girl if she had liked the movie. Then, to make things more awkward, he addressed that scene with her directly: "I wondered if maybe that wasn't a little old for you." Of course, now she was on the spot in front of a room full of people, but that was par for the course with Trump. As he always did, he wished everyone a good night immediately after the movie, and with his signature "Have fun, kids," dismissed all of us to head back to Shangri-La, Camp David's bar, which served drinks and snacks all night and had a great arcade, as well as an area to purchase Camp David items.

What surprised me the most about our movie nights was

that the president, who could never sit still for anything without talking on the phone, sending a tweet, or flipping through TV channels, sat enthralled whenever a movie was on the screen. It was one of the few times he seemed to be relaxed and quiet. Maybe it was because the movies I watched with him happened to be ones he liked. Or maybe it was because he saw the world as a movie, of which he was usually the star, and the language of film was the language that spoke to him best.

5

Trump Abroad

*Travel is about the gorgeous feeling of
teetering in the unknown.*

—ANTHONY BOURDAIN

In the spring of 2017, the White House announced that the president and first lady would take their first foreign trip—to five countries over eight days from May 20 to May 27, visiting Saudi Arabia, Israel, Italy, Vatican City, and Belgium.

It was a major event, an opportunity for the new president to meet his counterparts around the world and set a new tone for how he viewed the United States' role in the world. His "America first" policies were already ticking off and troubling a number of foreign leaders. The trip was a doozy, overly ambitious in terms of places and schedule and filled with complicated logistics. We never took one quite like that again.

Traveling abroad with the first lady, I quickly came to appreciate her view of life. Self-care is deeply important to her. She believed that relaxation was central to one's beauty regimen, as were, of course, spa treatments and facials. She thought it important to be surrounded by beautiful things and to feel comfortable.

Of course, she had the money and facilities all over the country to make that happen. Not everyone did. Once she boarded the plane for a long trip or had down time in a hotel, for example, she immediately changed into a luxurious robe and slippers. (I now completely agree with this particular concept, by the way, and wear a robe whenever I can.)

The president, however, seemed antsy on flights. He seldom changed out of his usual suit and tie, although sometimes he'd re-move the tie. He almost never slept on the plane, instead reading papers, doing work, visiting with staff, or watching TV. Needless to say, the first lady did not have that problem. Sleep, too, was a criti-cal component of her self-care routine. When she was at the White House, we almost never expected to hear from her before 10:00 a.m.

Foreign travel was always going to be a strain on the new first couple. The Trumps are homebodies, and they like their creature comforts. The entire time we were in Washington, the only place they ate outside the White House, as far as I can remember, was the Trump Hotel. The president had a few go-to meals that rarely varied regardless of where he was: well-done steak, cheeseburger and fries, or spaghetti and meatballs (for dessert, two scoops of vanilla ice cream). Mrs. Trump's diet was a bit more varied and much healthier. She loved soups, French bread, orange chicken, and fish dishes. Complicated, elaborate foreign dishes were gener-ally not going to be to their liking.

As I had never been to any of those countries, I was excited and felt privileged to be along for the ride. I thought there would be no better way to experience it than via Air Force One and in the safety of motorcades and secure hotels.

THERE IS A TON of work to be done anytime the first family leaves the grounds of the White House, so as you can imagine, an international trip requires an incredible amount of work and logistical co-ordination. Well ahead of the trip, the presidential advance teams are on the ground, working and negotiating with their foreign counterparts, as are the Secret Service and the various nations' security teams to ensure that every precaution is taken to safeguard the president and his delegation. Some countries are notoriously difficult when it comes to security, and there are a couple that do not allow the Secret Service to bring guns into the country, so tensions can sometimes get high before we even arrive.

Back in Washington, our staff worked with the West Wing to coordinate schedules so that Mrs. Trump could accompany the president to the appropriate events but also have her own solo engagements. Mrs. Trump, an extremely organized planner, didn't love all the last-minute changes requested by the president's team. She complained that they felt sloppy and unorganized, even though half the time it was because of the president himself. In the years ahead, foreign leaders such as Israeli prime minister Benjamin Netanyahu and French president Emmanuel Macron would learn that all they had to do was call the president directly if they didn't like the negotiations that had been done ahead of time by the advance teams. In order to ensure that the president would attend certain events, they'd make the requests personally, and President Trump would almost always agree. Every time that happened, it caused chaos for the operations, military, and Secret Service teams, but the president didn't care about the packed schedule or the lack of assets such as cars and helicopters—until we got there. Then he would see all the stuff he had randomly agreed to and how complicated it was and complain that the schedule was "inhumane."

IT WAS DURING THE first trip that I became acquainted with the notion of a "look book." The first lady's look book, filled with drawings and mock-ups of the outfits she would wear and how they would be accessorized, was the creation of a man named Hervé Pierre, Mrs. Trump's New York–based stylist and an extremely talented designer. Educated in Paris, including at the Sorbonne, Hervé had gone on to a glittering career in high fashion, working with and learning from some of the world's best designers, including Oscar de la Renta, Vera Wang, and Carolina Herrera. Through his work at other fashion houses, he had dressed the previous three first ladies as well, but he and Mrs. Trump had a long-standing relationship, and working with her was special to him.

Hervé would meet with Mrs. Trump in New York City or at the White House maybe once a month and go over her schedule. He would have clothes with him that she would try on and he would style, and they would decide on looks for all of her events. Even better, he would send us sketches or photos of each look that included the date of travel and event at the top, then the details of the clothing and accessories underneath. Hence, the look book became a big part of my life. Hervé would also add notes that included where designers were from and other options if she changed her mind about something. The look book became very helpful for me professionally, as the first question I generally got from the press was "Who is she wearing?" The sketches themselves were works of art, and I still marvel at how much time he must have put into them. A ton of attention to detail went into the looks, including what colors are offensive in certain countries, if there were designers from the countries who would be appropriate to wear, the protocol in terms of covering certain body parts, and on and on.

Mrs. Trump was very much into the look book; it was another relic of the world she had worked in and understood. "She knows what she likes," Hervé once said, describing his process of styling Mrs. Trump. "Our conversations were, and are, very easy," he went on. "She knows about fashion, as a former model. She is aware about constructions, so we have already the same vocabulary when it comes to designing a dress."

Hervé was an absolute joy to work with. He was everything I thought a high-end clothing designer would be and more, always so happy and positive, and very discreet. He and I hit it off immediately, and I will forever be grateful for all that he taught me without ever passing judgment on just how little I knew in the beginning. I recognized maybe 30 percent of the names that the first lady wore and had no clue about styling outfits and how to pair things. Hervé and I had a sort of real-life *The Devil Wears Prada* rapport, the talented designer taking this hopelessly frumpy fashion misfit under his wing. I shudder when I look at some of my earlier photos next to Mrs. Trump before I took "High Fashion 101" with Hervé Pierre.

WHEN YOU STEP OFF Air Force One in a foreign country, the chaos is almost too much to describe, but I'm going to try. As staffers, you are jet-lagged and scrambling to gather all your things so you can quickly exit down the back stairs of the plane, collect any necessary credentials, and locate your vehicle in the motorcade, all before the president and first lady get into the presidential limousine, aka "the Beast." If you were on team FLOTUS, you also had to ensure that her handbag made it into their limo and that all of her garment bags were safely off-loaded. While all of that was going on, advance people were yelling at you to hurry to the motorcade while a welcoming ceremony was taking place at the bottom of the main stairs that most of us stopped to take pictures of, despite all the yelling at us to hurry. Keep in mind that that was all happening after we'd likely slept very little, and on the floor of Air Force One. Though the plane is incredible, the sleeping arrangements for staff aren't great, and it is every man for himself.

What I remember about touching down in Saudi Arabia in May 2017 was the dust and the way Mrs. Trump looked when she emerged from the plane. It is my very favorite outfit that she ever wore—and that is saying a lot. Whereas the president was wearing his usual dark suit, white shirt, and overlong mono-colored tie, she had selected a black Stella McCartney jumpsuit and a huge Yves Saint Laurent gold belt that took my breath away. It was perfect in that she was sending a message: respectful, elegant, but also demanding to be seen. In a country where women are hidden from the public eye, that was an important message to send.

The motorcade kicked up dust and sand along the streets of Riyadh, with photos of POTUS and King Salman bin Abdulaziz lining the highway every few feet. Our first stop was at the King Abdulaziz Conference Center for a formal meet and greet of sorts. The inside of the building was ornate, with crystal chandeliers hanging from brilliantly painted ceilings. There were tall columns and majestic archways everywhere. It's right next to the Riyadh Ritz-Carlton, which, just a few months after our visit, would become known as the "gilded prison" where Prince Mohammed bin Salman imprisoned hundreds of rich and once powerful Saudis in a purge to flex his muscle.

But today the room was filled with men in the traditional Saudi outfit worn by royals and those high up in politics, all hoping to meet the president. The white robes are called *thobes*, and the red-and-white-checkered headscarf is called a *shemagh*.

Everyone in the delegation had a designated seat. I was placed next to a member of the king's council, with whom I had a frank conversation about women's rights and his wives. Yes, he had several wives, he said, but he added with great pride that each of them had her own personality and identity; I suppose it was nice of him to notice. And he said that the country was granting more liberties to women, which he supported. I looked at him as he said that with an expression that was captured in a White House photo by an Associated Press reporter. It was a look of "I don't know if I believe you, dude." But he seemed sincere, and I wanted to believe him. Still, what did I know? Saudi Arabia is a savage place for women.

It was a surreal feeling to sit there and have that conversation. I genuinely wanted to learn and understand some of the Saudis' ideals, while being careful not to be too open and to remember all the protocol rules (don't show the bottom of my left shoe, ensure that my scarf covered as much of me as possible).

When the president and first lady walked into the room, I was still awestruck by her appearance. This time, though, she was in a room full of very powerful Saudi men, and she was the one who commanded the room—a vision of an independent American woman that must have captured every man's attention. She and the president took their seats at the head of the room with the king and prince. Then I noticed something that caught me by surprise: Jared and Ivanka took seats there, too. Not only did it seem odd, it felt wildly inappropriate. I knew that Jared had some close ties to the royal family, but I didn't know the details. If they were up there as staff, where was the rest of the senior staff, most notably the chief of staff? If they were up there as the president's kids, had the rest of the Trump brood been invited? I highly doubted that. It was my first direct glimpse of what would become the constant issue of "Javanka" blurring the lines between staff and family and wanting whatever suited them best at the time. It seemed to me that whenever it suited her, Ivanka wanted to be treated as a senior staff expert on whatever issue caught her attention and resented being dismissed as the president's daughter. At other times, she'd want just the opposite.

AFTER HER POSITION AS a White House adviser was made official rather than informal at the end of March 2017, Ivanka sat down with Gayle King on CBS to make it clear that she wanted to be taken seriously. She told King, "I wanted to understand where I could be an asset to the administration, about how I could help my father and ultimately the country." She even named a few specific issues that she intended to focus on. "I'll continue the advocacy work that I was doing in the private sector," she said, "advocating for the economic empowerment of women. I'm very focused on the role of education."

But then she threaded the needle once again. "I'm still my father's daughter," she said with a smile, adding "I'll weigh in with my father on the issues I feel strongly about." The fine line between "asset to the administration and ultimately the country" and "my father's daughter" was the blurry one we were all constantly trying to figure out. In other words, she wanted to have it both ways. When there were issues that burnished her image, she was involved as official staff. When she wanted to distance herself from any of the president's controversies, she was just his daughter.

There were times when Ivanka would go complain to "Father"— what she called the president in the Oval Office—when she wasn't getting what she wanted. All the adult kids referred to him that way—"My father said this" or "as I was telling my father." Aside from the protocol issues, it exhibited a complete disregard for the president and first lady—as if no one should matter but them.

I knew Mrs. Trump well enough by now to see that she was irritated by that but not surprised. It was the beginning of an irregular cycle in which Mrs. Trump would get riled up by Javanka's presumption and want us to fight it and then other times when she would shrug her shoulders, say, "What can you do? Nothing will change" and tell us not to bother. In that part of the trip, it was the latter. Jared was the one who had what appeared to be an oddly close relationship with the Saudis, presumably due to his family's business ties, so Melania knew that it would be pointless to complain.

The president was oblivious to the appearance of maneuvering between his wife and daughter, or so he led us all to believe. In all the time I worked with the president—and we talked about everything under the sun—he never once indicated any awareness of tensions between Ivanka and Melania. I did learn that every once in a while the president and first lady would argue about the fact, as he put it to her, that "you don't like my kids." But my impression was that it was a rare occurrence. My own theory was that he was in denial most of the time. Both of the women, Melania and Ivanka, were important to him, and it wasn't in his interest to take one side or the other. So he pretended that there was nothing to see, and they were smart enough not to press him too hard. The Melania-Ivanka dynamic was just a fact of life, and I would see it play out again and again. And again.

THE STAFF'S ROOMS WERE paid for by the Saudi government, and we were told we could order anything we wanted on the room service menu. It was the first and only time that happened, because we came to find out that it was against the rules. I don't know if the state department ended up paying for staff rooms, but I imagine they did. At the time we didn't know, or maybe didn't care, that it wasn't allowed and that any gifts given to us had to be reported and turned in so they could be assessed for value before we were given the option to purchase them. I have no idea what the end result was or if true, but I remember some people talking about the fact that Jared and his staff had been gifted Rolex watches, ornamental swords, and various textiles.

One day on the trip, Mrs. Trump went on a school visit and I ended up getting in a big fight with a Saudi cameraman. He didn't like a female—me—telling him where to go, so he stormed out. I also tried walking outside alone at some point and quickly learned that was not a good idea. When I stepped outside our hotel, intending to walk across the parking lot to the convention center, I was surrounded by five men in uniform pointing at me to go back inside. I didn't understand what was going on, but fortunately one of our Secret Service agents saw me and escorted me to where I needed to be. Those two episodes gave me a real appreciation of some of the liberties that we enjoy in the United States.

Later in the day, the president addressed the Gulf Cooperation Council. That was the setting for my first small act of rebellion on behalf of my girl. I went out to see the stage and where Mrs. Trump was supposed to sit so that I could brief her appropriately. I immediately saw that Ivanka was positioned to sit behind and to the left of the president, who would be standing at a podium, Jared was to be next to Ivanka, and Mrs. Trump would be off to the right. I could tell immediately that Ivanka would be in the camera shot more than the first lady would. I had no idea if an advance staffer had been told that that was where Ivanka wanted to be or if it had all been done innocently. But whatever. Javanka had inserted themselves more than plenty on the trip and should have been sitting in the audience with the rest of the senior staff anyway. So I quickly walked onto the stage and moved the name cards around.

After making what I thought to be a sly little move, I went into the hold room to let Mrs. Trump in on it. As Jared and Ivanka were both there, I asked the first lady to come with me to the bathroom. I urged her into a stall with me, which she seemed to find a little curious at first, and who wouldn't? Then lowering my voice I told her what I had done with the name cards and where I wanted her to go once she went out to the stage. As she listened to my efforts to outwit Ivanka, she smiled, not so much because of what I had done to Ivanka—as I say, she knew Ivanka was just a fact of her life—but because she had staff who looked out for her and understood what was appropriate in such situations. She didn't say much other than confirming that she was to sit in "the seat to the right, yes?" and that was that.

I went back out to make sure that the seating cards didn't get re-arranged. As I was standing watch, John McEntee, the president's body man, approached me. John had been with the president from the beginning and was loyal to the very, very end. He is some-one I considered a friend but who would change drastically in our last year at the White House. He was tall, dark, and handsome, "right out of central casting" as the boss would say. "Do you have any face powder?" he asked me before going on to say that the president felt he looked "shiny" and for some reason John didn't have the usual arsenal of products with him. Many pundits, com-menting on Trump's extremely tanned (or orange) appearance, assumed that there was a tanning bed in the White House. There wasn't, and I would know. The president's look was created with makeup that he put on his face every morning, as if he were going to be appearing on a TV show. Which, in a sense, he was.

I wasn't sure how I felt about the president using my makeup and immediately worried that the compact didn't look clean in some way. I grabbed my powder out of my bag but explained to John that I had obviously been using it, figuring that that would make it a no-go for our germophobe president. But John just smiled. "It'll be fine, you're good looking," he said, a bit oddly. Not going to lie, it thrilled me that during one of his most important speeches up to that point in the administration, the president of the United States was wearing *my* makeup.

WE LEFT SAUDI ARABIA for Israel on May 22 and upon arrival were greeted with an extraordinary welcome ceremony, one that the president would talk about for the next four years. There was a red carpet that went from the bottom of the Air Force One stairs, where President Reuven Rivlin, Prime Minister Netanyahu, and others were waiting to greet the Trumps, all the way to a beautiful stage. Along the way were members of the Israeli military holding flags. At the bottom of the stairs, everyone faced the flags as the US and Israeli national anthems were played. Then the band started playing military marches as the president walked, first with Rivlin and then with Netanyahu, toward the stage for welcome speeches. It had all the pomp and circumstance that the president loved and the first lady appreciated.

That was also where the infamous "hand slap" occurred, causing endless speculation about the state of the Trump marriage. At one point, while they were walking together down the red carpet, President Trump was a few steps ahead of Mrs. Trump. He reached out and slightly behind him with his left hand, apparently to try to take Mrs. Trump's right hand. Without breaking her stride and with a flick of the wrist, she appeared to swat his hand away and kept walking. *Haaretz*, an Israeli newspaper, tweeted out a video clip of the incident zoomed in and in slow motion, as if it were replaying the John F. Kennedy assassination footage. The rest of the world's media loved it. "Melania Trump and the Hand-Graze Seen Round the World" ran the headline in *Vanity Fair*. The *Independent* in Great Britain came up with its own explanation: "Melania Trump Slaps President's Hand Away to Say She Won't Be Treated as a Child, Says Body Language Expert."

Since the beginning of the administration, any number of the president's critics had been convinced that Melania Trump was basically a hostage of her husband. They assumed that the attractive foreign-born model had married Trump for his money and held him in contempt as much as they did. Endless memes sprouted up to support the hypothesis, including a memorable one at the inauguration in which Mrs. Trump appeared to be glowering at the president as soon as his back was turned to her. The "hand slap" was another piece of "evidence" that Melania hated Donald and couldn't even stand the thought of being touched by him. "Free Melania" became a popular slogan on social media—with the hope that once Melania was released from whatever devil's bargain had tied her to her husband, she would speak out publicly against him.

Alas, the truth wasn't as interesting. For those who care to know, Mrs. Trump "slapped" her husband's hand away that day because she thought it was against protocol to hold hands at such a formal ceremony. Melania was a rule follower, sometimes to a fault, and her husband knew that. He often tried to hold her hand or messed with her hands on purpose in front of the cameras to irritate her. I found it amusing, but the press never seemed to like that explanation, even when I offered it. The "Free Melania" narrative could not be undercut so easily.

A KEY STOP ON the trip was our visit to the Church of the Holy Sepulchre followed by the Western Wall, also known as the Wailing Wall. It was an incredible experience. The Wailing Wall is thought to date back to about 20 B.C. and is located on the western edge of Judaism's holiest site, the Temple Mount—hence "Western Wall." The two sides of the wall are separated by sex, so POTUS and FLOTUS approached their respective sides at the same time and placed their rolled-up handwritten blessings deep into the wall per tradition. Because I assumed she was worried about her English, the first lady had actually asked me to write her blessing earlier on Air Force One.

Unfortunately, that day it seemed Mrs. Trump wasn't happy that Ivanka was also at the wall and had inevitably made a display of being there. Melania never really went into long diatribes if something upset her. It would be more short questions accompanied by eye rolls: "Did you see she was at the wall with us? Always there. It is not appropriate." Then she would wait for my reply to let her know she was right to be irritated and that it was inappropriate. I respect that Ivanka wanted to be there because she was Jewish, but really she would have been anywhere as long as TV cameras were present.

At that point on the trip, I was starting to get as fed up with her behavior as the rest of the East Wing team was. Ivanka was constantly getting into the press shots that truly should have been reserved for the president and first lady. It was yet another example of the Kushners putting themselves on the same level as the first couple, and it was unseemly. For Mrs. Trump, it was about protocol and the rules; for all of us as staff, it was about allowing her to be in her role and have the people of the United States see her representing them with dignity and class. I've no doubt that that day was an important and emotional experience for the Kushners, but they didn't need to make the visit alongside the president and first lady while the cameras were rolling. They could—and should—have had their moment at the wall in private. Mrs. Trump seemed relieved to see that I shared her frustrations with her daughter-in-law, so much so that she eventually let me in on the nickname she had privately given her: "the Princess." Many times after that I would hear one of her favorite stock complaints, "Princess always runs to her father."

We also let the first lady in on our own nicknames for Jared and Ivanka. It seemed they had been fashioning themselves as experts on everything—Jared was taking charge of dealings with Mexico and building the wall, and Ivanka was suddenly an authority on women's rights and empowerment around the world—but they really didn't have a clue what they were doing (just like the rest of us, I might add). Because they dabbled in a bit of everything and could be precocious and self-absorbed, we in the East Wing dubbed them "the interns," and the nickname stuck. Mrs. Trump was amused and herself used the nickname every now and then.

ON MAY 23, WE left Israel and headed to Rome, where the president and first lady would have an audience with Pope Francis. Being Catholic, Mrs. Trump was very excited about that portion of the trip and had chosen her outfit with great care—a black-and-white Dolce & Gabbana single-breasted coat with a matching dress and pumps. On her head she wore a D&G Chantilly lace mantilla.

An audience with the pope is a rare and special occasion, one not taken lightly by the many Catholics on our team. Understandably, only a few people from our entourage would be allowed into the Vatican, let alone the room with the pope. Sean Spicer, who was not only a senior staff member but a devout Catholic, was hopeful that he would be allowed in. But it was not to be. Instead, in what was now a running slideshow in my life, Jared and Ivanka, "the interns," neither of whom was Catholic, took his place while Spicer sat in a van outside. That one really pissed me off. Sean is a man of deep faith, and he deserved that special moment. He didn't complain, though, just asked that someone take his rosary inside to be blessed. Was he pissed at Jared and the Princess? Maybe. He should've been. But he is a gracious guy and seemed to be in a precarious position with the president by then so probably couldn't afford to make enemies, especially not those two. Overall, the visit was a success and even included a lighthearted exchange between Pope Francis and Mrs. Trump. He asked her if she fed the president potica, a well-known dessert in Slovenia. Mrs. Trump wasn't sure what he had said. She thought the Holy Father asked if she fed her husband pizza, which by the way she would never do, but after a few rounds with the translator they all had a good laugh at it.

After the visit with the pope, the president and first lady parted ways for solo events of their own. The first lady and our team headed to the Bambino Gesù Children's Hospital, the first children's hospital in Italy, which in 1924 was donated to the Holy See and became the hospital of the pope. We visited a room full of children with various ailments, and Mrs. Trump colored with them and even spoke with them in Italian, one of her five languages (English, Slovenian, French, German, and Italian). We also visited the hospital's version of an ICU, a stop that had a bigger impact than we could ever have imagined. Only Mrs. Trump and I were allowed into the room with the medical staff, and once inside, I understood why. The kids were extremely ill, and many of them would likely not survive for long. The room had roughly a dozen children in it, and Mrs. Trump went to each one and spent time with them while learning about their conditions. It was at that moment that I saw Mrs. Trump's authentic warmth and compassion for children. She was always great with kids, and each visit was sincere, but I could see that this one was having a profound effect on her—although I found it strange that she never once looked at me.

The last patient she visited was a young boy with a chronic heart condition. His prognosis was not good, and he and his family were waiting desperately for a donor. The boy was maybe ten years old, but he looked like he was five. His motor skills were labored, and he didn't speak. Mrs. Trump sat with him for quite a while. She read him a story, talked to him, made silly faces, and held his hands. After hearing about his condition and the likely outcome from his doctors, my eyes welled up with tears. But after watching Mrs. Trump with him I almost lost it and stepped out of the room to compose myself. Later that day Mrs. Trump told me that she had seen that I was getting emotional. "I saw you were about to cry, so I had to stop looking at you because I knew I would cry," she said. "And we needed to be strong for the children." That was probably the only time Mrs. Trump ever spoke with me about her capacity to feel deep emotion, and I was both surprised and relieved to know she had it in her.

After the visit, as we were on our way to Brussels, our entire team spoke of the hospital and that part of the visit. It seemed that each of us had seen or met someone who had impacted us in some way. After we landed, we found out that the little boy she had visited with for so long had gotten a donated heart. I truly believed it was a miracle. Mrs. Trump had been in the presence of the pope, then with that small child, who had been waiting for so long and didn't have much time left, and suddenly he had that heart. To this day, I remember that as one of the most profound and wonderful moments during my time in the White House.

AFTER OUR VISIT TO Brussels, we headed to the G7 summit in Taormina, Sicily. It is one of my favorite places I have ever been. It is small and quaint with cobblestoned streets, adorable shops, and the best part: impeccable Italian food. A mountainside town with breathtaking views of the ocean, it made me understand how people can call a geographic location romantic.

On Friday, May 26, 2017, I had my first brush with a PR disaster. The spouses' program that day included a visit to Elephants Palace in Catania. I knew what Mrs. Trump planned to wear that day from Hervé's trusty look book. It included a sketch of a coat, along with a photograph. Alongside that Hervé had put a description: "Dolce & Gabbana all over embroidered cocoon coat with multi-colored silk flowers over a D&G vanilla floral jacquard dress with matching vanilla fabric pumps and purse." The description and Hervé's sketch were a lot more appealing to me than the photo of the actual coat, which to me looked like a jacket that my grandmother would wear. But as per usual, when I actually saw Melania Trump wearing it, she was stunning.

For our helicopter ride to the palace, Mrs. Trump removed the jacket because we were all crowded inside and she didn't want to crush any of the flowers. It briefly ended up on my lap, and I was shocked at how heavy the damn thing was. My novice fashion brain wondered if it was even comfortable to wear and how she kept from sweating profusely in it. The things one does for fashion.

We were greeted by a cheering crowd, and Mrs. Trump flashed them all a smile as she walked inside the building, that heavy-ass flowered coat "casually" hanging over her shoulders. While Mrs. Trump and the spouses took part in a reception followed by lunch, I was in an outer room with other members of staff. As I fended off the aggressive flirtations of a "gentleman" from France, I started getting calls from the press. They were all about the cost of the heavy jacket. Apparently the thing had cost $51,500, and after all the visits and trips we had been on, including to the little boy who'd received a heart transplant, that was the big story.

"First lady Melania Trump wore the equivalent of an average American family's income on her back for her appearances Friday in Sicily, Italy," scolded USA Today. CNN obsessively reported that it was "by far the most expensive item of clothing Trump has worn this entire trip." At least CNN got the price tag right. Most other outlets just lazily rounded down to $51,000. But Robin Givhan, the fashion writer at the *Washington Post*, actually stuck up for Mrs. Trump. "Clothes can be deeply symbolic," she pointed out. "And Trump's choice of Dolce & Gabbana—an Italian brand that has been deeply inspired by Sicilian culture—for a trip to Sicily makes sense." "And frankly," she added, "the floral coat is beautiful."

First of all, I had no idea that that crazy heavy and presumably hot coat that had touched my lap had cost as much as some people's homes. Worse, all the good work and interesting things Mrs. Trump had done on that trip might as well have never happened at all. Those were the moments when, for all the mistakes we'd made, it felt like we just couldn't win. It's not as if Michelle Obama and Laura Bush walked around in burlap sacks. I can't imagine that their wardrobes drowned out all other coverage, as Mrs. Trump's did. Givhan did note in her article that Michelle Obama had once worn $540 shoes to a food bank and pointed out, "There's a big price gap between a pair of designer sneakers and a coat that costs as much as a house in some parts of the country, but the fundamental point is the same: fashion shame."

Knocking the first lady for her clothes was an easy, lazy news story that always won attention on the networks, which were eager for embarrassing material on the Trumps to boost their ratings. Of course, there would be days when I'd look back fondly at that wardrobe malfunction, as things got even worse.

6

———

Chief Two

One must be strict, even in the little things.

—JAMES BOSWELL

On July 28, 2017, President Trump tweeted that General John Kelly, who had been serving as secretary of homeland security, had just been made White House chief of staff. As for the outgoing chief of staff, the president wrote, "I would like to thank Reince Priebus for his service and dedication to his country. We accomplished a lot together and I am proud of him!"

The reality was a little harsher. We had just returned from Long Island, where the president had given a speech, and had landed back in Washington. I was sitting in the motorcade, in the staff van, when the tweet went out and then became the hot topic of gossip among the rest of us, the survivors of one of the first Trump purges. I watched Reince walk off the presidential plane, get into his car, and be driven away with no fanfare at all. So much for that. He had been White House chief of staff, one of the most coveted and powerful jobs in Washington, for about six months. The president deplaned next and went under the wing to praise all the work that Reince had done, then moved quickly to how wonderful and strong John Kelly would be. "John Kelly will do a fantastic job," he said. "He has been a star, done an incredible job thus far, respected by everybody, a great, great American." I had known the president long enough by then to think, *Yeah, we'll see how long this lasts.* I did wonder if General Kelly had any misgivings about the way Reince had essentially been fired by tweet. Like so many others, I imagine he never thought it could happen to him.

I DIDN'T KNOW MUCH about General Kelly at that time, but it had been painful to watch Reince being dragged through the press wringer and slowly losing Trump's confidence as air might leak from a tire. So we all welcomed a change in leadership. And to be honest, the "Trump people" saw the firing as an opportunity to get rid of the "RNC people." The remaining RNC people, many of them blindsided by Reince's casual firing, were feeling nervous and rightly so.

In fact, Sean Spicer, a Reince favorite, had already resigned, seemingly over the news that Javanka had arranged to bring Anthony Scaramucci in as communications director after judging that the press coverage in "their father's" White House wasn't good enough. That plan worked out well, guys. "The Mooch" lasted all of eleven days before giving a long-winded, curse-filled off-the-record/on-the-record conversation with a reporter from the *New Yorker* that got him fired. Like a shooting star, the Mooch shone brightly for an instant and then was gone, to be replaced by Hope Hicks, who wisely always kept the word "interim" in her title. Filling Sean's role was my old friend and colleague in the press shop, Sarah Huckabee Sanders.

So for those keeping score, not seven months into the Trump administration, we'd already had two chiefs of staff, two homeland security secretaries, three communications directors, two press secretaries, and a partridge in a pear tree.

GENERAL KELLY WALKED INTO the White House just a few weeks before one of our first huge embarrassing, insulting, tone-deaf disasters. In August 2017, after a violent confrontation in Charlottesville, Virginia, between supporters and opponents of the removal of Confederate statues, Trump had issued a statement condemning violence "on both sides." The problem was that the violence had predominantly come from the pro-Confederacy side, which included white supremacists and other hateful bigots. Trump's comments, which he doubled down on later during remarks at Trump Tower—a scene that had General Kelly dropping his face into his hands—played right into the narrative that Trump was a bigot who winked and nodded toward the David Dukes of the world. At one point during the controversy, he seemed to call those in the pro-Confederacy crowd "very fine people." It was a dumb thing for him to say, and then he got too stubborn to try to clean it up. This was a pattern that would prove even more disastrous for him as time went on. By contrast, Mrs. Trump saw the violence in Charlottesville for what it was—vicious racism—and she decided quickly to issue a statement condemning it. She also did not like the "very fine people" comments at all and tried to show her husband by example what he should do. It was too bad she didn't do that more often.

AS THE DAYS WENT by, the gossip mill started to churn about General Kelly. The Trump White House was growing notorious for people jockeying for position, and things were no different under Kelly. What piqued my interest the most was the rumor that he had made a deal with the president before taking the job. Allegedly, he had demanded that among other things, he have full control of who entered the Oval Office—including Ivanka and Jared—and the president had agreed. As with everything, hindsight is 20/20, and it would become clear well after he was gone that he had done the best job at reining in the "crazy" and keeping the Oval Office and access to the president well guarded. But holding back the crazy in the Trump White House was like one of the last lines in the Al Wilson song "The Snake" that the president was so fond of reciting at his rallies: "You knew damn well I was a snake before you took me in."

Kelly was a hardened military guy, and it showed. On one trip to China in November 2017, I saw that firsthand at a dinner that the president and first lady attended. Chinese government security is very intense, and it kept track of where every person was and didn't allow much staff around—that always involved a huge negotiation before the trip even began. That evening, the military aide who carries the nuclear football got separated from the president. The separation was only a few feet, but that was still not supposed to happen. A bit of a melee unfolded.

Kelly and a few others calmly tried to explain the situation to the Chinese military. Then they started yelling. When that didn't work, the military aide, along with Ronny Jackson, the president's doctor, a couple of Secret Service agents, and General Kelly got tangled up—literally and figuratively—with the Chinese security. They pushed and shoved their way past the guys with guns while the president and first lady sat in a room a few feet away, oblivious. General Kelly was right in the thick of it and got knocked around a good bit (he gave as good as he got) until the Chinese security guys backed down.

I walked into the hotel's executive lounge late that night

looking for a snack and found General Kelly's main Secret Service agent there, feeling horrible that he had been waiting in the car while the whole thing went down. The agent was huge, formidable-looking, but the sweetest and kindest teddy bear on the inside. I remember him saying "I failed tonight, Stephanie. I should have been there for the general. It is all my fault." To which I replied that that was silly and we both knew that General Kelly had probably enjoyed himself. What better story could an old marine tell than fighting Communists to save the world from a potential nuclear holocaust? I'm being dramatic, but still. A couple of years later, that agent passed away from cancer. The president, first lady, Secret Service agents, and many colleagues in the White House attended his service across the street to pay honor to him and his beautiful family.

THE RUMORS ABOUT GENERAL Kelly's deal with the president seemed true at first. As soon as he began his job as chief of staff, he cut off Jared and Ivanka's free rein in the White House and made them observe, at least for a time, proper procedures and channels.

Jared was the kind of guy who thought he had to have an answer for everything. I think it was mostly because his father-in-law expected him to. So he started to dip into meetings on whatever topics that caught the president's attention whether he knew anything about the subject or not. For example, take the president's most famous promise, building the border wall with Mexico. There were constant conflicts within the administration over the wall. We couldn't even agree on the facts. I once sat in a meeting with Jared, Stephen Miller, and officials from various agencies, including the Department of Homeland Security, Immigration and Customs Enforcement, and Customs and Border Protection. Jared was determined to get everyone to agree on the language about how many miles of border wall had been built—a seemingly simple task. The problem was that it depended on how the wall was defined. Did we mean brand-new border wall? Or existing wall that had been refurbished? And what about existing fencing? That was one of the times when I saw how Miller could drive people away. He was a speechwriter basically dictating to experienced professionals what the facts were going to be and what they were supposed to say. Meanwhile, it seemed that Jared just wanted an easy fix. He'd say something like "So we're all going to say we've built four hundred miles of the wall." When someone objected, he'd say, "Don't worry. I'll make it happen." Then he would be off on some other issue he was suddenly an authority on.

General Kelly objected to that kind of operation and worked to stop Jared's freelancing whenever he could. Jared didn't like that. In fact, he became extremely vocal and bitter about it and spoke about it until the very end of the administration, starting many sentences with "In the Kelly days . . ." before grumbling about one thing or another. The two guys were polar opposites: a grizzled, hardened, older military guy who had suffered great loss in service to the country versus a thin whiz kid who got by on looks and family connections. They were destined to collide. That would be a big problem—big for Jared and even bigger for Kelly.

THE SUMMER OF 2017 also gave me another piece of the puzzle that was the marriage of Donald and Melania Trump. During our first trip to France, I was standing mostly alone with the first couple as they prepared to go out and speak to US Embassy employees and their families. While they waited to go on, they watched the proceedings on a television screen in front of them. Mrs. Trump, dressed in a striking red dress, leaned in to her husband, who whispered into her ear. A few moments later I saw them kiss each other. All of that was quite unusual; it was in fact the only time I ever saw them express any physical intimacy in public. Neither believed in public displays of affection, and Mrs. Trump, being Eastern European, preferred kisses on the cheek. I was so surprised that I took a picture of the scene on my phone. Like everyone else, I had no idea what was really going on between those two. But for one moment they looked like a real couple with real affection for each other.

Maybe they were finally getting their groove as a first couple, I don't know. On the same trip, Mrs. Trump, unusually playful, started to twirl around the room in that red dress, holding up one side of her skirt and pretending she was in *The Sound of Music.*

Something made the Trumps seem closer than ever. Of course, it wasn't long before they were pulled apart again.

7

Our Storm

When a storm is coming, all other birds seek shelter.
The eagle alone avoids the storm by flying above it.

—UNKNOWN

In private, he called her "Horse Face." That was his explanation, at least in his mind, for why he was innocent of the charge made by Stephanie Clifford, aka porn star Stormy Daniels, that the two had had an affair.

"She has a horse face," he said over and over again. "I wouldn't touch that."

At one point during the scandal, Trump even ordered someone in the press office to share the nickname with reporters. Right before he was to walk to Marine One, he ordered the person to go to the press people waiting on the South Lawn and tell them the nickname. "Go out there!" he ordered. "Go out there and tell them she's a horse face."

There was no way that was going to happen. So instead the person did what we all tried to do when Trump wanted us to say something insane to the press: pretend it wasn't happening and hope the president would forget. There were a lot of times like that.

*   *   *

I WILL NEVER FORGET the day in early January 2018 when I first got a call from a reporter giving me a heads-up that the *Wall Street Journal* would be reporting that an adult entertainment star named Stormy Daniels had accepted a secret payment to stay quiet about an affair she claimed to have had with the president in 2006.

Making the subsequent phone call to Mrs. Trump truly sucked. I'd been cheated on before and knew what it felt like. I couldn't imagine how it would feel if that kind of betrayal—true or not— was front-page news all over the world.

I walked into her empty (as usual) office in the East Wing for extra privacy since it was such a personal issue. Her office was in a corner space with no offices except mine immediately outside, and I wanted a place with complete discretion. Then I called her personal cell phone.

It was awkward, and I didn't know how to approach it or lay it out delicately. So I stuck to giving her the plain facts as I knew them. "Ma'am, I got a call from a reporter that the *Wall Street Journal* is going to do a story on payments supposedly made to a woman named Stormy Daniels to keep quiet about some alleged affair with the president," I told her. "I am being asked if we want to comment on this."

She didn't say much, just "Okay." And then "Don't replay." I was amazed at how calm she was. I assumed that she was embarrassed and probably mad, but she definitely did not seem sad. What struck me most was that she didn't seem surprised. I think she'd known who she was marrying, but the fact that the affair had supposedly happened right after she gave birth to their son probably stung. In addition, their son's name was going to be in all of the stories, which meant that the eleven-year-old would surely see it himself or be told about it by his friends. To be honest, I think the country knew what her husband was, too. It was pretty common knowledge that Trump had committed adultery before. In previous eras, an extramarital affair by a sitting president would have all but doomed an administration; it led to Bill Clinton's impeachment, for example. With Trump, it was just another day at the office.

I was pissed off on her behalf as a woman and a mom. Over the last several months, I had come to really like and admire her. Now she would have to put on a brave face for her son and have to stand next to her husband and smile at all sorts of events. I was mad at the president, and I was mad at the people around him who in my opinion had to know about it, such as Michael Cohen, Dan Scavino, and maybe even Hope. There wasn't much Trump did that that crew didn't know about.

I was expecting her to do with me what I would have done. Get pissed. Yell. Say something like "That asshole, how could he do this to me?" I would go into every detail and complain and vent to all my girlfriends about how I'd been wronged. I think it was ego on my part to expect that of her—my presumption that she owed it to me as a friend, when I was really just staff. But if she did have those feelings—if she was mad as hell—she didn't share it with me. Not at that moment, at least. I would see flashes of that later, as the story just never seemed to die.

The alleged affair had been in the news before but hadn't received much attention until the *Wall Street Journal* published its story on January 12 under the headline "Trump Lawyer Arranged $130,000 Payment for Adult-Film Star's Silence." That same day, the president and first lady, as well as their traveling staff, left for Mar-a-Lago for the weekend. Generally on our weekend trips to Florida she kept to herself during the days, spending time with her son and parents, spa visits, naps, some work, but she almost always had dinner with the president on the patio at night. That weekend, she skipped dinner both evenings and was very quiet with me, so I knew something was up.

Our office went into lockdown mode. She wanted to know about every press inquiry I received but again told me I was not to respond to any of them. She didn't talk about it much other than when she'd ask, "What's new?," meaning what was in the news, and I had to keep telling her it was all about the alleged affair. As her comms director, I offered one suggestion—that she let me say something to the effect that the first lady was "focusing on being a mother and to give her family some privacy." But she didn't even want to say that. "That wouldn't be good enough for them," she said of the press. "They would want more and make more things up." She wanted to wait and think about it, oblivious, or not caring, that the press would speculate about what she was doing and thinking anyway.

OVER THE NEXT FEW days, the drip-drip of Daniels revelations continued. On January 19, a week after the *Journal* story broke, *InTouch* published a full transcript of an interview with Daniels it had been holding on to since 2011. The gems from that one ranged from the bizarre ("He said that he thought that if he cut his hair or changed it, that he would lose his power and his wealth.") to the eyebrow raising ("He told me once that I was someone to be reckoned with, beautiful and smart just like his daughter.") to the plain gross ("I can definitely describe his junk perfectly, if I ever have to."). Unfortunately, she apparently could and later did.

But the most gut-wrenching bit from the interview, for me, was the single mention of Mrs. Trump and their son. When Daniels was joking about Trump's hair, he told her, "Yeah, my wife even did my son's hair like that, as a joke." Daniels apparently said something like "Yeah, what about your wife?" to which Trump replied, "Oh, don't worry about her" and in Daniels's telling "quickly,

quickly changed the subject." Sharing an intimate family moment, such as playing with your son's hair, with your side chick—and then dismissing your wife with "don't worry about her"—yeah, that part was hard to read.

The day after the full *InTouch* interview dropped was January 20, 2018, the one-year anniversary of the inauguration. As we worked together on a tweet to mark the occasion, I caught a glimpse of the real Melania Trump—the pissed-off spouse and human being. As usual, I wrote the tweet for her, which included the POTUS Twitter handle. She edited it to take out the reference to her husband. Instead her tweet simply said, "This has been a year filled with many wonderful moments. I've enjoyed the people I've been lucky enough to meet throughout our great country & the world!" Not a single mention of her husband, whose inauguration we were ostensibly marking.

She also tweeted out a photo of herself on the arm of a handsome military aide—and I remember thinking "Okay, it is on." Finally I thought I was seeing her anger, albeit in a passive-aggressive and still private way. For all those who thought, or continue to think, that the relationship between Donald and Melania Trump is purely transactional, you didn't see her when the Stormy Daniels saga went on and on. Whatever the real relationship is between these two—and they are the only ones who truly know—there is clearly an emotional tie. I felt that Mrs. Trump was embarrassed, and that she wanted him to feel embarrassed, too. Whether he is capable of that or not, I don't know.

I wasn't the only one who noticed her slight of her husband. The media inquiries started to come in, and once again I was told, "Don't replay." January 22 was their wedding anniversary, and the timing couldn't have been more awful. We also canceled her trip with the president to Davos, Switzerland, for the World Economic Forum, which caused further speculation. The trip had been in flux even for the president, and the simple truth was that she was tired of waiting for a final schedule, so she canceled. We didn't have solo events planned anyway, and she didn't like being on trips with nothing to do. Once again, that fueled the rumors that she was "furious" with her husband, which I don't think she minded.

SOON AFTER, WE HEADED to Mar-a-Lago for a long weekend. On the way there, we talked about all the news and speculation, and I reiterated to her that if we wanted the press to calm down, staying silent was not the way to go.

"Look, I think we should put something out to feed the beast," I told her. I again suggested saying that she was focused on her role as first lady and mother. Period. I thought reminding people that this was a family issue with a child involved might make reporters feel a little more sheepish about poking around.

But she reiterated what she'd told me before: "No, because that will never be enough."

I said, "You have to give them something."

"No, because then they will just attack me."

Over and over we went through the dance. And we never came to a satisfactory resolution.

I felt so bad for her and wanted to comfort her as a friend. So that weekend, I asked her if she wanted to join me for a walk on the beach. I had no idea if that was an inappropriate suggestion from someone on her staff, but she rarely left her suite unless it was to go to the spa or dinner on the patio, and I had always thought that was such a waste when she had a semiprivate beach only steps away. Anytime I was at Mar-a-Lago, I made it a point to run on the beach in the mornings and walk or lie out whenever I could—the sun, the sand, the waves, and all the good people watching were good for the soul. I sent her pictures of the ocean whenever I was out, too; it felt like such a shame for her not to experience the beautiful surroundings.

I was shocked when she responded, "Maybe" to my suggestion of a walk. I immediately wondered what she would wear. What I thought I was conveying with my invitation was this: Let's hang out woman to woman. You are going through some shit. Let me help you get it off your chest, or at least you can breathe fresh air and stare at an amazing ocean under the sun.

That was not how she took it at all. She called me and asked, "And there will be photographers? Paparazzi?" She assumed that I saw it as a press opportunity, not a chance for us to hang as peeps. I felt like such an ass to have offered.

I was torn. I didn't like the idea because either she'd be photographed walking alone, which would fuel headlines, or she'd be walking on the beach with a staffer, which would be weird. But I also really wanted her to get out to that beach! She seemed to like the idea of a photographer, though, so I got on the phone with a still photographer I trusted to ask how something like that would work. I remembered that when Michelle Obama had gone to a Target, a photographer had been placed inside to capture her being a "real person," so I thought it would be easy. As usual, I was wrong. The photographer said he couldn't do any kind of exclusive and I'd have to offer the same opportunity to the other media outlets that were generally in the pool. I then moved on to a tabloid publication, which said that it would do it but couldn't agree that no one would jump out of the bushes and ask her questions. The entire idea was scrapped, and my plan for the woman to simply take a walk on the beach with me for the sole reason that it was a great day and the sunshine can be healing was for naught.

Later that night, though, I felt a small victory when she allowed me to tweet—finally—that speculation and gossip about their marriage were inappropriate and she was focused on her role as mom and first lady.

I had initially written "wife, mother and First Lady," but she nixed the "wife" part, so I assumed she was still pissed off.

Either way, in my own mind, I felt that that weekend was a turning point in our relationship. Strange as it may seem, I felt she had let me in just by suggesting that we put a photographer on the beach, and she had (kind of) taken my advice on communications. She even accepted my own take on things when I told her, "If this were me, I'd be having a hard time right now." She just shrugged, though. "This is Donald's problem," she told me. "He got himself into this mess. He can fix it by himself." I marveled at that kind of attitude, still do. Because it is completely correct—when a person cheats, it is his or her own problem, nothing that the other person did. On the way home I felt she let me in again when she told me that she wanted to go to the upcoming State of the Union Address separately from her husband. That was another unorthodox move. But I knew not to ask any questions or put up any fight. I had a feeling she knew it would cause a stir and fan the flames of speculation—and I think she liked that. So I stayed quiet and knew we'd need to make it happen.

I HUDDLED WITH THE team when we returned, and we decided that since the guests at the State of the Union would technically be sitting in the first lady's box, it would be best to host a reception of sorts for them in the Diplomatic Room of the residence. We would then put the guests into her motorcade to get them to the Capitol early so everyone could get settled. It was a win-win situation, because the guests would be given special treatment that was well deserved and Mrs. Trump could leave ahead of the president without causing too many headlines. The Trump children had already begun creating issues with regard to where they would sit and insisting that they all be in the president's motorcade, so that seemed a perfect solution for our team. One of the kids was always late, and Mrs. Trump didn't want to deal with that headache, either.

The best part for me came a few hours before we were supposed to leave, when Mrs. Trump called to let me know that she wanted one of our military aides to escort her throughout the Capitol because "the floors were so slippery."

I laughed to myself because I'd seen the woman navigate dirt roads in her heels. Those famous slippery Capitol floors! But I said, "No problem, I'll get to work." Her chief of staff and I then spent the next hour walking around the military office discreetly "shopping" for a good-looking aide.

You would think that would have been enough drama to have to handle for one evening, but no. Jared and Ivanka decided last minute that they wanted to join our motorcade so they could arrive at the Capitol early, too, which in my opinion really meant to get as much camera time as possible. None of us was happy about it, including Mrs. Trump, but we obliged, and Lindsay Reynolds and I made a plan: the minute the motorcade stopped, she and I would run, sprint, whatever it took to get between Javanka and Mrs. Trump. That way we could let Mrs. Trump enter with her military aide and not be forced to share the camera shot. It was clumsy and very obvious, but we managed to get in front of Javanka at the last possible moment and then planted our bodies there, pretending we had no idea they were behind us. Their Secret Service agents eventually pushed us aside, but by that time we had accomplished our self-appointed mission. The State of the Union was the first time Mrs. Trump had been seen publicly since all of the stories had hit, so naturally everyone wanted to see if her face was puffy from crying or if she'd be wearing a FUCK YOU, DONALD T-shirt. I'm sure that many were disappointed when she walked down the steps in an all-white pantsuit looking tan and rested and wearing a big smile. Always the poised model. Always made me proud.

THE DRAMA OVER TRUMP'S "other women" never really went away. On February 16, Ronan Farrow of the *New Yorker* published a story that focused on "catch and kill," which is a common practice in which a tabloid pays someone for the exclusive rights to a story and then never publishes it. Farrow talked about Karen McDougal, a former *Playboy* model, who had allegedly fallen victim to the practice after she'd claimed to have had a long-term affair with President Trump from 2006 to 2007. Dylan Howard, a top executive of American Media, the parent company of the *National Enquirer*, had allegedly paid McDougal $150,000 for her story in August 2016 and then buried it. If true, it was a classic catch and kill.

Once again, the news hit on a Friday, when we were planning to leave for Florida. I hadn't even called Mrs. Trump to speak to her about the story before I received a call from her to let me know that she wanted to drive to Air Force One ahead of her husband. She surprised me, saying "I do not want to be like Hillary Clinton, do you understand what I mean? She walked to Marine One holding the hands with her husband after Monica news and it did not look good," referring to Monica Lewinsky. I didn't argue. Though the comms person in me knew that it would once again cause headlines and speculation, the previously betrayed woman in me was right there with her. I wouldn't want to put on a show for the world, either, and nor did she it seemed.

The weeks that followed included mostly news and interviews regarding Daniels and the hush money payment by the president's lawyer Michael Cohen. Mrs. Trump told me that she and the president had been on the phone with Cohen at one point and Michael had said it was all his doing and President Trump knew nothing. That was when he was still defending Trump. He had apparently told them both on speakerphone that he knew the allegations weren't true but he knew how the media would react and was genuinely concerned that it would affect the outcome of the upcoming election.

It was interesting for me to watch Mrs. Trump's evolution at that time. She went from staying completely silent to acting as if maybe she didn't believe any of it to finally telling me that despite all the denials by the president, she had a feeling all of it was true. And she would say about Cohen's version of events at that time, "Oh, please, are you kidding me? I don't believe any of that bullshit," and it made me love her more and more. I am just speculating, but it seemed that she was more angry at being what she perceived as humiliated in the press than at the news itself. At other times she would blame Michael Cohen for having made the payments; I guess he seemed like an easier person to direct her anger at.

But after the Karen McDougal interview with Anderson Cooper on CNN on March 22, her mood shifted. After that, she finally seemed to be genuinely angry with her husband and no longer hid it. We were at Mar-a-Lago when it aired; it was a Thursday evening. I was in my room watching it, and the president and first lady were having dinner on the patio—a conversation I was glad to be out of. I knew instinctively, both as a woman and as her comms person, that that interview would be harder to watch and hear about than everything Stormy and her sleazy lawyer, Michael Avenatti, had said combined.

I think most of the women who are reading this will understand. Though cheating under any circumstances is awful and hurts like hell, cheating one time with an "adult film actress" who showed zero empathy or remorse would be easier to handle than a longer-term affair with someone who seemed genuinely sorry for what had happened and had apparently fallen for a married man who she thought really loved her too. Again, there is no "good" affair—but maybe some are easier to understand than others.

That night, after the McDougal interview had aired, Mrs. Trump sent me a text: "did you watch?" I told her I had and offered a few thoughts on the interview: "She seemed like a nice person, remorseful, and will get far more sympathy than Daniels." She replied with an eye roll emoji, then added her usual "I recorded it. Will watch later." She seemed to have a lot more patience and self-discipline than I did, that's for sure.

We stayed behind when the president left on Sunday, not necessarily because she was angry but because it was the family tradition to stay all week at Mar-a-Lago for their son's spring break. I was so relieved that we wouldn't be traveling back with everyone else. The rest of the staff kept asking me what was going on and what she was thinking, and I refused to answer anything other than "She's fine." Not that she asked for or needed it, but I was more protective of her in that period of time than I had ever been before or would be after. Though she always told me, "This is his problem, he created it," I suspected on a very basic human level that it had to hurt, had to be embarrassing, and she must have been worried about what their son was seeing. Or maybe I was projecting, I don't know.

THE MOST BIZARRE MOMENT of the entire Stormy Daniels saga, for me at least, came some months later, in September, when the first excerpts from her tell-all book began to leak. She complained about the president's endowment, or lack thereof, saying he was "smaller than average" but "not freakishly small" (whatever that means) and that it had an "unusual" look, "like a toadstool."

The president called me from Air Force One. The funny and frustrating thing about the presidential plane is that despite all of the great technology at our disposal, calls from there still sound horrible, like someone talking through a walkie-talkie from far away. So the conversation was pretty brief. At some point, he brought up the Stormy interview. "Are you guys commenting on anything?" he asked. "You guys" meant his wife.

"No, sir," I replied. "I don't think we are."

He paused, and I sensed he was gauging whether I'd tell him how pissed his wife was. I didn't take the opportunity as my own opinion changed every day.

"Did you see what she said about me?" he said after a moment, referring to Stormy. Then he added, unsurprisingly, "All lies. All lies."

"Yes, sir."

Then I figured out exactly what he was concerned about. "Everything down there is fine," he said.

What the hell was I supposed to say to that? I kept it to a simple "Okay," praying that somehow we'd get disconnected.

"It's fine," he repeated.

"Uh, yes, sir," I replied. Well, that was awkward. Not in a million years had I been wondering about that. Not in two million years had I ever thought I'd have a conversation with the president of the United States about his penis. Thankfully the call ended shortly after that. And no, I never told Mrs. Trump about the call.

THERE WERE, OF COURSE, other women who made allegations against the president, including one made by a Florida volunteer for the Trump campaign. She claimed that the then candidate had made inappropriate comments and forced a kiss on her in a trailer full of people. I had actually been at that event, so I knew it hadn't happened.

Another was the writer E. Jean Carroll, who claimed that Trump had sexually assaulted her in the 1990s in a woman's dressing room at Bergdorf Goodman. Mrs. Trump sided with her husband on that one, finding many of the details of the story unbelievable. Regardless, the allegations did lead to another disturbing moment between me and the president.

I was in the Oval Office with Jared, Dan Scavino, and a few others when the Carroll claims came up. Trump was doing his usual routine when accused of misconduct: "She's a liar," "She's gross," "Do you think I'd be with that?," on and on, mostly attacking her physical appearance rather than the "I would never do that to my wife" line.

Then, at one point, sitting at the Resolute Desk, the president shot me an unusual look. It's hard to explain, but it was like he was staring me down or piercing into my soul.

"You just deny it," he said. "That's what you do in every situation. Right, Stephanie? You just deny it," he repeated, emphasizing the words.

It felt like a weird loyalty test. He knew how close I was to the first lady. I think he wanted to see how I would respond to that. He continued to stare at me. I was unsettled but replied, "Yes, sir." And for anyone who is wondering, no, that was another conversation I never shared with Mrs. Trump.

AFTER THE STORMY DANIELS story broke and all the allegations that followed from other women, I felt that Mrs. Trump was basically unleashed. She had always been independent from her husband, but now, as a wronged and publicly humiliated first lady, she seemed liberated to do whatever she wanted, or didn't want, to do. After all, what was he going to say? We wrote statements that contradicted the administration or dived into personal family affairs. We almost never ran anything past the West Wing, though I did try to give Sarah Sanders a heads-up whenever I could. It was freeing, really. We were rebels, Thelma and Louises lovin' life and livin' the dream, and answered to no one but Mrs. Trump.

She was also more willing to rebuke the president and his allies and didn't appear to give a damn what he or the West Wing as a whole thought about it. The first time Mrs. Trump had me publicly strike back at someone close to the president was in June 2018, and the target was none other than Rudy Giuliani.

Rudy had been at some speaking engagement when he was asked about Mrs. Trump and the allegations of her husband's affair with Stormy Daniels. Rudy replied with something like "She believes her husband, and she knows it's untrue." That . . . did not go over well. Number one, as we all know now, Mrs. Trump hates other people speaking for her. Number two, she sure as hell did not want Rudy Giuliani, a three-times-divorced aging man known in some circles as a lothario, delving into her personal life.

I thought Giuliani gave off weird vibes when he was around the president. It seemed he was always up to something—which, it turns out, he usually was. One thing I noticed was that although the president preferred having a small gaggle of advisers around him most of the time, he almost always met with Rudy alone. I saw that plenty of times, both at the White House and at Mar-a-Lago. On one occasion the entire bar area at Mar-a-Lago was blocked off so Rudy and the president could sit at a table in the middle of the room totally undisturbed—no staff or anything. We all stood around just outside the windows, as though we were looking at animals in a zoo. I thought it was a little odd, but frankly I never wanted to be in a Rudy meeting for fear of getting roped into whatever crazy scheme he may be peddling that day— and then being cross-examined about it later by some committee.

So, no, there's no way Rudy had anything close to an idea of what was going on in Mrs. Trump's head. Of course I started getting requests for a response from members of the press corps and informed the first lady. She didn't offer her usual "Don't replay." In fact, she did not hesitate to respond. She told me, in no uncertain terms, what I was supposed to say, which was the following: "I don't believe Mrs. Trump has discussed her thoughts on anything with Mr. Giuliani." I was stunned, proud, and mortified all rolled into one. Obviously that kind of thing is what the media salivates over, and I knew it was going to make waves. I also agreed with her sentiment wholeheartedly. I assumed that it wouldn't go over well with POTUS, but what was he going to do? He was already in hot water. I did my due diligence in advising her that it would most certainly make news, but she was resolute. She told me to send it. "Do it right now." I did just that and picked up the phone to call Sarah Sanders. I did hate such moments because I knew it would have to be up to Sarah to tell the president what I'd said— but that's part of the job. I think it was one of the times the first lady didn't mind someone speaking for her.

THE NEXT TIME OCCURRED a couple of months later. The "Be Best" campaign was one of Mrs. Trump's most memorable initiatives in the White House, but like much else we did it was a case of good intentions gone awry. That was one of the times when Mrs. Trump's language issues caused problems—they collided head-on with one of her worst qualities. To describe how Be Best came to be, you need to understand Mrs. Trump's stubborn streak and the multiple different people giving her advice. As I stated earlier, Mrs. Trump's adviser Stephanie Wolkoff felt very strongly about the importance of social and emotional wellness in children—a noble cause.

She spoke to Mrs. Trump about it often, and they both agreed that it is a critical component of a child's journey toward adulthood. Mrs. Trump had also said in an interview prior to her husband being elected that one of her initiatives would be working against cyberbullying. Of course, after he was elected, the media began trying to pin her down on that, given that her own husband was notorious for his vicious tweets. The Trump administration had also made the opioid epidemic a huge priority, and Kellyanne Conway, who headed up that effort, rightly felt that Mrs. Trump would be an asset in helping the country understand the devastating effects on babies born addicted to opioids.

All three of those issues are important, especially in today's world. But each one presented its own problems and challenges, so we tried to get her to focus on a single issue. For whatever reason, Mrs. Trump insisted on combining all three of them into one huge platform.

First, it was clear to everyone on the team that cyberbullying would be an issue that, even with our greatest efforts, would never go away. It also didn't make sense for her to own an issue on which we'd struggle to prove measurable progress. But Mrs. Trump felt she would be "attacked" if she didn't go through with what she had mentioned in the earlier interview. She was right, too, she would have been—but only for a short time rather than the entire four years we were in office. Second, the idea of social and emotional wellness was still fairly new, and it was a difficult concept to explain to the general public. Third, opioid abuse with a focus on neonatal abstinence syndrome was considered a "West Wing thing," and there were concerns that we would look as though we couldn't think of something on our own. Finally, the three issues, though certainly centered around helping children, were vastly different and would be very tough from a comms perspective to brand and put into short and succinct talking points that would be easily digestible to our main audiences: children and parents.

Mrs. Trump held her ground, though—she wanted to do all three, and it was our job to make that happen.

We went around and around on language and messaging—and I think that was where Stephanie Winston Wolkoff's issues became most intense. We finally settled on quasimessaging, which would include "three pillars," our reasoning being that there was no reason to choose any one thing because in today's world children have to deal with a myriad of issues, which is true. And we changed one of the pillars from "cyberbullying" to "online safety" for obvious reasons, not that it ever mattered. Next came the name and the logo. The team brainstormed dozens of different names and logos that were presented to Mrs. Trump. Some examples included "Children First," "Online Aware," and "Hope4Kids"—and all of them were shot down. Mrs. Trump came up with Be Best, and none of us was thrilled with it, because we knew it wasn't proper English. Wolkoff and I agreed on that point, and we both tried in our own ways to talk her out of it.

True to form, Mrs. Trump dug in, so I started thinking of ways I could incorporate Be Best into less awkward sentences. "Helping children to do all they can to Be Best in all that they do" became a mainstay talking point. Again, we sent many logos up—different colors, looks, all of it—and one day she sent us what would become the Be Best logo. She had drawn—written?—it herself using the pen feature on her iPhone, and her sister had helped her refine it a bit. She seemed to really love it, and we had been going at this for weeks, so we all threw our hands up and went for it. It was her initiative, after all.

I remember that when it came time to build the official Be Best website, our digital team came to us with all the things that were wrong with the logo, including every single thing we'd tried to explain to the first lady. Suffice it to say, we shut them down as quickly as she had shut us down.

*   *   *

AFTER BE BEST WAS announced, it became a weapon turned on us by the press all the time, especially the part about online safety, since, as reporters and pundits loved to point out, Mrs. Trump was married to the cyberbully in chief. And they weren't wrong.

The whole affair came to an unfortunate climax in August 2018, when the president tweeted a harsh attack on LeBron James, questioning his intellect. James had just been on Don Lemon's show on CNN saying that he would not sit down across from the president if he were given the opportunity. The president commented, "Lebron James was just interviewed by the dumbest man on television, Don Lemon. He made Lebron look smart, which isn't easy to do." He also made sure to note "I like Mike," implying that LeBron would never be as good as Michael Jordan. I'm not going to wade into that comparison.

The interview itself had been about a new charter school that LeBron was opening in Ohio, so naturally, as soon as the president tweeted, I was inundated with reporters asking if the first lady agreed with that statement and if the president was violating the concept of the Be Best initiative with that kind of language. It was a pet peeve for me and a catch-22 for our office. Mrs. Trump could not control her husband and truly had only the best intentions for children. But nobody seemed to care about that.

The first lady and I went back and forth on that one. On the one hand, LeBron was opening a school that served underprivileged children, which was completely in line with Be Best. On the other, the only news that would come out of commenting would be that she disagreed with her husband. We decided not to comment at first, but then the pressure from the press grew. "I'm being killed, it's everywhere" she told me eventually. So we came up with a statement that in my eyes focused solely on the good work of the school being opened and answered the question of whether she would ever visit it. "It looks like LeBron James is working to do good things on behalf of our next generation and just as she always has, the First Lady encourages everyone to have an open dialogue about issues facing children today. As you know, Mrs. Trump has traveled the country and world talking to children about their well-being, healthy living, and the importance of responsible online behavior with her Be Best initiative. Her platform centers around visiting organizations, hospitals and schools, and she would be open to visiting the I Promise School in Akron."

I might have thought we'd put out a cleverly calibrated statement, but the press didn't. The headlines focused on Mrs. Trump rebuking her husband. *US Weekly* screamed, "Melania Trump Sides with Lebron James After Husband Donald Trump's Tweet Slamming Him." The next weekend at Bedminster, I was sitting at a table outside with Hope Hicks and Hogan Gidley when the president came back from a game of golf. He spotted me sitting there and called me over. "Did you put that LeBron statement out?" he snapped. "Why would you do that?"

Sheepish and a bit scared, I said, "Yes, sir, I discussed it with Mrs. Trump, and we thought that was the best way to go since we were getting so many press inquiries asking if she would ever visit."

"I didn't like it," he said with annoyance and clear anger. "Not at all." He had to know his wife approved it, of course, but he sure as hell wasn't going to yell at her. Then he stared at me for a minute before walking over to where I had been sitting. He proceeded to go out of his way to heap over-the-top praise on Hogan and Hope, everyone but me, the weak problem child. After he walked inside the club to grab his lunch, I sent Mrs. Trump a text telling her about the exchange and saying "He is really mad at me."

She replied, "Yes, he did not like statement. It is ok, be strong" followed by the arm muscle emoji. "Be strong" along with an emoji of some kind was her advice to me for pretty much everything. Easy for her to say.

The president didn't acknowledge me for a few days, and that sucked. There was nothing worse than when you were on the receiving end of his ire. It was the first time he was really upset with me, and I felt guilty for days, butterflies in the pit of my stomach.

Then he got mad at me again—in an even more memorable fashion.

8

The Damn Jacket

This is a day without a trace of reason.

—FROM "CITIZEN" BY BROKEN BELLS

Well, fuck. No other way to say it. What a shitty day this turned out to be.

Mrs. Trump had been genuinely upset and concerned about what she was seeing on TV about the Trump administration's sweeping immigration reform plans, including a policy that involved separating children from their parents when they entered the United States illegally. The flash point was the disturbing images of children in facilities that looked like cages—never mind that the same facilities had been established and in use during the previous administration. This is a problem the Biden administration is also confronting, with no slam-dunk easy solution. I want to be clear that this part of the book is not meant to be political. Immigration is a hot-button topic, and people seem to think it's cut and dried—but it's not. I worked for the attorney general in Arizona and lived in a border state for almost twenty years, so I know that illegal immigration is a tough issue and will be difficult for any administration, no matter what side of the aisle it's on, to solve.

As the family separations increased, they received a whole new level of coverage, and Mrs. Trump noticed it immediately. She monitored the news daily, so she always knew what was going on, and she was very unhappy with the images of children crying or being taken from their parents. Morally and politically, it didn't sit well with her.

After some discussion over the implications it would have in the West Wing, we toyed with the idea of flying to Texas and Arizona so that Mrs. Trump could see things for herself. I (once again) called Sarah Sanders to give her a heads-up that that was something we were considering. Sarah rightfully expressed concern about how such a visit would look, especially if it suggested to the public that Mrs. Trump was demonstrating opposition to her husband's "zero tolerance" illegal immigration stance.

Of course I explained all of that to Mrs. Trump. "I understand this," she replied. "But I want to see for myself how the children are being cared for." She let me know that it would be fine "with Donald."

I also wondered if she was digging in because the West Wing people were expressing concern. She was adamant that we ran things separately from the other side of the White House and that she was her own independent person. She made it clear to us all the time that we did not answer to the West Wing. That wasn't how a first lady normally behaved—usually she took great pains not to show any divergence from her husband's policies in public—but we had left "normal" behind a long time ago. At that point I just went with it. So to be clear: what turned out to be a disastrous next few days began with Mrs. Trump's well-meaning intention to offer support for the separated children, whether it reflected poorly on her husband or not.

The night before we left, someone from the advance team called to say that one of the facilities we planned to visit was having a head lice outbreak. Though the people there were doing everything they could to clean the facility and treat the children, was the visit something we still wanted to do? Like her husband, the first lady is a germophobe. I figured she would use the head lice outbreak as an excuse to get out of the trip. As previously noted, she didn't like to leave the White House anyway.

To my surprise, she didn't change her mind. "Let us still go," she said.

I was not only astonished but privately very proud. It really was *that* important to her.

GENERALLY WHEN WE WENT on a trip, I would wait in the Diplomatic Room on the ground floor of the Executive Residence to greet her and catch her up on anything pressing before the motorcade departed. But on that particular day, I was already in the motorcade, on a conference call with the advance teams in McAllen, Texas, preparing for our visits, which meant that I was caught completely by surprise by what happened next.

After the first lady's chief of staff got into the van and I was off the phone, she said something like "We're wearing a jacket that says 'I don't care' on it." At the time, I was working on three things at once, and frankly I didn't think much about it. She definitely didn't mention that the phrase was in big, bold white letters on the back. Or that it was the jacket from hell that would ruin our entire day. And it certainly wasn't in any "look book" I had seen. It turned out that it was a jacket from Zara that Mrs. Trump had ordered online all by herself. I believe that Hervé even put out a statement saying that he knew not where it had come from. Hervé was a smart and extremely talented professional—he wasn't going down for that.

I did send Mrs. Trump a text message as we were en route to

Joint Base Andrews, reminding her that the press would be under the wing before we took off.

She responded with "oh really? I thought only when we landed?" And I reminded her of an email I'd sent her the night before listing all the press movements. She replied "ok," and that was that.

When we arrived at the plane, she got out and walked up the stairs. She was in white jeans and a green jacket, but the jacket was cinched in at the waist, so I still didn't notice anything. But the press did. Someone snapped the jacket in its full glory, and the photo was already making its way around the world.

UPON ARRIVAL IN TEXAS, Mrs. Trump had changed into a different outfit, which was not uncommon for her. We visited a detention center in McAllen that provided education, medical attention, food, and shelter for the children. Mrs. Trump asked a lot of questions of the staff and the children—including if they were able to speak to their parents, and they said they were. All of us were pleasantly surprised at how well the facilities were run under tough circumstances and what seemed to be the best of intentions with regard to those caring for the children. Of course, the authorities had known that the first lady and the press were coming, so they had obviously put their best foot forward. Our visit didn't mean that there weren't serious problems elsewhere. Still, Mrs. Trump wanted to show that she was concerned about the children, and she clearly was. But no one was going to care about any of that.

It wasn't until we were heading back to Washington that a reporter sent me a text saying "You know wearing a jacket that says 'I don't care, do you?' on the back is going to be a story, do you have any comment?"

I responded by asking what he meant, and he sent me a picture of the jacket. When I saw it, I cringed. Of all the many, many articles of clothing Mrs. Trump had available to her, all the great designs Hervé had put together, she'd decided to travel to border states where kids were being held in detention wearing a grass-colored jacket that in bright white letters read, I REALLY DON'T CARE, DO U?

What a stupid thing to do. That was what I thought, at least, but that wasn't what I said to the press, and still in that moment I thought he was wrong or there had to be some explanation. I told the reporter that a piece of clothing was a silly thing for anyone to focus on, and he replied, "While that may normally be the case, this is still going to be a big deal." I protested again: if she didn't care about the kids, which was the obvious suggestion, she clearly wouldn't have gone to Texas in the first place, and I noted that she changed her clothes for the actual visit with kids. Then I went to take a look at the damn jacket. Maybe there was something I was missing.

I walked into her cabin on the plane and asked, "Does your jacket say something on the back? Because I am starting to get inquiries."

At that moment the first lady had a look on her face, the kind you might have when you know you fucked something up. She stared at me for a minute and then showed me the jacket, which was thrown over the captain's chair in her cabin. When I saw the message on the back, a range of emotions went through me, from guilt for not looking into it sooner to anger at her making such a rookie mistake.

I asked her why she had worn it, and she said, "It's just a jacket." I truly do not believe she was sending a message that she didn't care about the kids. Melania Trump can be accused of many things, but she is a mom who is devoted to her son and to helping children. I don't think she meant to slight anyone else's children in

such a cruel way. To be honest, I don't know what she was thinking. Maybe she thought the press would never see it because she
had forgotten they'd be under the wing of the plane. Maybe she
was sending a message to someone else by wearing that jacket,
but I was never sure. Maybe she got up late and grabbed a jacket
quickly, knowing she was going to change on the plane. I truly
don't know the answer. In any event, a well-meaning excursion
had turned into another major PR disaster for the Trump White
House. As if we needed it. Part of my mind had an image of Ivanka
and Jared cackling at us from wherever they were at the time.

We went back and forth for the next fifteen minutes on what to
do. At one point Mrs. Trump asked if we should put a circle and
cross through the "don't" portion of the jacket, as if the reporters had gotten it wrong. That was a creative suggestion, at least,
something the president himself might have come up with, but
a terrible idea. Everyone had seen the jacket already. Reporters
would surely compare photos from when she had first boarded
the plane. Also, might I add that covering it up, tampering with
the evidence, so to speak, would make it an even bigger story. She
kept pointing out that she had changed outfits for the visit itself
and the jacket was just her casual traveling outfit. I'll say again
in her defense (I'm still doing that, I guess) that she had clearly
forgotten that the press would be covering her departure from
Washington. But that wasn't their problem.

As we were talking, inquiries from all the major networks and
print publications were pouring in, and I told her we needed to
give a response before getting off the plane. I suggested the truth
(at least in my mind): that she hadn't realized what the jacket
said. But she didn't like that suggestion, because, I think, to her
it would have looked somehow disorganized. We settled on "It's
just a jacket," and I headed back to my seat to start responding. I
did suggest that she not wear the jacket off the plane or even tie it

around her waist. But she refused, saying "That will make it seem I did something wrong." That, of course, was only minutes after she'd wanted to doctor the jacket itself. Man, she and her husband could be very similar sometimes!

Once we arrived, I stood on the tarmac, watching her go down the stairs and get into her car, my heart sinking, feeling like the worst press person in political history, all while that message—I REALLY DON'T CARE, DO U?—was right in front of the traveling press pool, all of whom were gleefully and frantically taking pictures. The drive from Joint Base Andrews to the White House was about twenty minutes in the first lady's motorcade, and during that time I watched Twitter explode.

All the headlines focused, of course, on the jacket rather than on the purpose of her trip to the border in the first place. "Melania Trump Wore a Jacket Saying 'I Really Don't Care' on Her Way to Texas Shelters," the *New York Times* told the world. CNN made sure that everyone got the timing straight: "Melania Dons Jacket Saying 'I Really Don't Care. Do U?' Ahead of Her Border Visit—and Afterward." Geraldo Rivera waded into personnel matters and said our chief of staff "should be fired for allowing her to wear that jacket to the border." NPR got out the rhyming dictionary—"This Jacket Caused a Racket"—and asked, "What, Exactly, Does Melania Trump Not Care About?"

SHE DIDN'T CARE ABOUT the media frenzy over her jacket, that's for sure. Or at least she pretended not to care; she wore studied indifference as though it were armor. But I did care, because it was my job and, at best, I looked totally incompetent. I was devastated that a trip begun with such good intentions was being completely overtaken by a jacket. I'd like to think that had I seen her wearing that jacket in the morning, I would have persuaded her to take it off. But I'll never know. The truth was that it was such a stupid mistake, exacerbated by her stubbornness. And it was about to get worse. Someone else was watching the disaster unfold in real time. And he wasn't happy.

As soon as we pulled onto the South Lawn of the White House, we were told that the president wanted to see his wife in the Oval Office. Uh-oh. He had never summoned her like that before, at least not in front of staff. It had a feeling of being called to the principal's office.

Lindsay and I stood in the Palm Room, the room on the other side of the colonnade, opposite the Oval Office. I think we were both planning to send the boss off with a "good luck" pat on the ass and then head back to the safety of our offices in the East Wing. No such luck. The first lady said, "You are coming, right?" She was clearly looking for backup, which only made me more nervous. She wasn't someone who ever looked for backup, especially when it came to her husband.

Anyway, what choice did we have? "Of course," we said. We are coming with you. Wouldn't miss it. Sure to be a great time!

As we walked into the outer Oval Office, Madeleine Westerhout, the president's executive assistant, gave us a look. We were in big trouble. As if we didn't know that already. Madeleine was a master at telling you what was happening without saying a word.

Mrs. Trump, Lindsay, and I walked into the president's private dining room, where he was seated at his usual chair at the end of the table, watching TV. Dan Scavino was also present, sitting to the president's immediate left and facing us as we walked in.

The boss was in a sour mood, which of course I could appreciate. He looked at his wife and then at us with annoyance. The first words out of the president's mouth were "What the hell were you thinking?" and I am not sure who of us he was talking to.

To my surprise, Mrs. Trump sat down in the chair next to him on the right (still wearing that damn jacket, by the way) and smiled. That was weird. Husband Trump didn't know quite what to do with that, but it seemed to soften him a little bit.

He looked at me. "Why did you let her wear that jacket?" His tone was half angry, half genuinely confused. Even he knew there had to be some stupid explanation behind it. It was almost as if he was marveling at how any group of people could fuck up something as badly as we had.

Well, now, what the hell was I supposed to do? Throw the first lady under the bus to the president of the United States? Pin it on Hervé, who was totally innocent? Obviously not, but I had no explanation, so I went with a little humor and simply replied, "Sir, I bought the cardigan I am wearing right now at Target. Do you think she is going to take fashion advice from me?" I have no idea if the man even knows what Target is, but it seemed to work. He didn't press any further. He of all people knew how stubborn his wife could be and that there was likely nothing any of us could have done to stop her even if we had tried. He had also been a bit more sheepish toward his wife since the Stormy Daniels controversy of earlier that year, so he could push only so far.

We sat there for a few minutes until Trump came up with a thought. "You just tell them you were talking to the fucking press," he said. "Let me tweet that. Dan, take this down. Do it now. Ready?" He then dictated a tweet to Scavino that she had worn the jacket to send a message to all the media and haters out there.

That was damage control—and it didn't seem to matter where the explanation came from. In the middle of a firestorm, I didn't stop to question it. Working for Trump, we were in a constant battle for survival against his opponents, against investigations, and against most of the press. This was our story now, and it didn't really matter if it wasn't true. Casual dishonesty filtered through the White House as though it were in the air-conditioning system.

All five of us wordsmithed the tweet for a bit. I continued to be surprised that Mrs. Trump was not saying much and even more surprised that she was going to allow him to speak for her. That suggested to me that although she wouldn't admit it, she knew she had messed up royally and that she was responsible for all of it. For the first and only time that I ever saw, Mrs. Trump put herself at the mercy of the POTUS Twitter account.

Scavino sent the tweet out that read: " 'I REALLY DON'T CARE, DO U?' written on the back of Melania's jacket, refers to the Fake News Media. Melania has learned how dishonest they are, and she truly no longer cares!" Then we got up to leave.

Trump seemed very pleased with himself and was now in a great mood. He had, in his mind, come up with the perfect cover story. As we were leaving the Oval Office, he yelled after us, "I just saved your ass." Mrs. Trump laughed, said, "Okay, Donald," and continued on her way. And she was still wearing the jacket! I wish I could have taken it, thrown it into the fire, and sprinkled the ashes far away. I do wonder if that thing will go into any kind of a museum one day, not gonna lie.

SO THERE YOU HAVE it, the story of the jacket and why we responded the way we did. Believe me when I say it was tough to have people saying that her staff didn't handle her properly and should have never let her wear the jacket. Or that I'd lied when I'd said it was "just a jacket."

I stood just off camera later on when the first lady participated in her first and only sit-down interview as first lady, with Tom Llamas of ABC News.

"Let's talk about the jacket," Llamas said.

"The jacket," the first lady responded. We knew what was coming.

When Llamas asked why she had worn it, Mrs. Trump stuck to the story that the president had come up with. "It's obvious I didn't wear the jacket for the children," she said. "I wore the jacket to go on the plane and off the plane. And it was for the people and for the"—she paused to find the phrase since it was not one that came naturally to her—"left-wing media who are criticizing me. And I want to show them that I don't care. You could criticize, whatever you want to say, but it will not stop me to do what I feel is right."

Llamas's last question about the jacket hit me in the gut, though. "Your office released a statement during this time saying the jacket is just a jacket," he reminded her, using my exact words. "So you *were* sending a message with the jacket?"

"It"—she paused again, as if she were trying to remember our story—"it was kind of a message, yes. I would prefer that they would focus on what I do and on my initiatives than what I wear."

So she completely contradicted what I had said to the media. But that is the job of the spokesperson: take the lumps in the press, and do all you can to make your principal look good, even if it means that you look inept or simply as though you're a liar.

9

Africa

*You cannot change the past, but you can
always change your perspective.*

—WAYNE DYER

Since she refrained from participating in several of the events
that a first lady usually did and remained mostly a mystery to
the rest of the country, Mrs. Trump's first solo international trip
was a big deal. Her chosen destination was probably unexpected.

We chose to visit Africa in the fall of 2018 for many reasons. One
of them was to try to soften public perceptions. The comments
her husband had reportedly once made about African countries—
calling them "shithole" countries—seemed to motivate the first
lady to show that not everyone in the Trump family felt that way.

When we started to talk about where we would visit, she outlined some conditions: "I do not want it to be a place that is only
pretty; this is not a vacation. I do not want to just do fancy tea or
go to nice dinners." She also told us that she wanted to go somewhere no other first lady had been. Finding a place was not an
easy task. When we did identify a potential location, there was
ultimately an issue or two or three. Either we couldn't find chil-

dren's programs to visit, many hospitals were not feasible, or the Secret Service had an issue with security. We wanted the trip to be safe but also not look like a luxury trip on the taxpayers' dime. Mrs. Trump also wanted to have her days packed with activities so she wouldn't be criticized for wasting taxpayer money. After a few weeks, we decided on Africa and put together a potential schedule for her to look over and approve.

THE CONTINENT OF AFRICA is vast, and as in North America, each country has its own set of problems and issues. We did not seek West Wing or National Security Council approval, but we finally settled on Ghana, Malawi, Kenya, and Egypt for the four stops. Working with the United States Agency for International Development (USAID), we planned for each stop to have cultural, educational, and diplomatic components, while trying to promote the pillars of Be Best. She had also agreed to do her first-ever solo sit-down interview as first lady with ABC News, so that was going to be a big deal. She had a very specific look in mind in terms of the background for the interview, so we decided to do it in Kenya prior to a visit to an elephant orphanage and safari tour.

Our first stop was in Ghana, which was probably the most profoundly somber stop on our trip. There we visited the Cape Coast Castle, one of around forty "slave castles" that had served as prisons for slaves en route to the Americas. We were greeted by a tour guide, who walked Mrs. Trump through many rooms and told stories of how the slave trade had begun. We were shown a room that had held hundreds of men and women, with tiny windows that barely let in light. There was a small ditch dug down the middle of the room, maybe six inches deep and wide, and it was explained to us that it had been used as a bathroom. Each room was horrible, and the tour guide was brilliant in the way he told us the

grim and heartbreaking story of the way the people kept there
had lived. And when I say brilliant, I mean that he told it in a way
that we almost lived it—we felt their pain, their misery, almost
understood what it must have been like to be treated as cattle. The
thought that human beings were held in such horrific conditions
until they were placed on ships in the middle of the night, only
to live in even worse conditions until they arrived at their destination, was hard to stomach. There were rooms for the women
that were equally as brutal. We stopped at an altar to pay tribute
to all those who had lost their lives and those who had lived under
such cruel circumstances. I remember feeling distinctly ashamed
that that had ever been allowed to happen and about our country's complicity, since so many of the people had been shipped
to the United States. Mrs. Trump felt deeply impacted as well. In
conversations later that day she said, "I did not know. The conditions were so horrible. Did you see the rooms? How can people do
that? Everyone should see these things, and we should talk about
it openly." We all, of course, knew about and abhorred slavery,
but were less familiar with all the details of its brutal origins. The
emotional visit concluded with Mrs. Trump walking through the
"door of no return," the door that the people had left through to
be loaded onto ships to be taken to the various countries that used
slave labor. The ride home from Cape Coast Castle was quiet. It
was a solemn day, and it was clear afterward that no one, including Mrs. Trump, the staff, the press, and even our Secret Service
detail, was in the mood to talk.

OUR VISIT TO MALAWI changed my life. It just did. It is a poor country but full of people who are so welcoming and do so much with
so little. To this day I tell everyone that I would move there if I
could. In fact, I told the president more than once that I wanted to

be an ambassador there, which confused him every time I said it. "I would peg you as a European girl or just somewhere more elegant," he'd reply. We had only one stop in Malawi, and that was to visit the Chipala Primary School. It was a school of roughly nine buildings, none of them with doors or windows. There were no desks in any of the buildings; instead, the children sat on the floor to learn. The school had so many children that there were not enough buildings, so many of the classes were held outside, with the children sitting in the red-brown dirt that got onto absolutely everything.

Mrs. Trump toured each classroom both inside and out and sat in on a few of the lessons. She spoke to some of the children, saying "Hello. How are you?" When the language barrier was too great, she offered high fives and hugs. At one point she sat in on an English lesson and repeated some of the basic words with the children, such as "apple," "cow," and "flowers." She even got in on a little soccer game with some of the children when she was handing out soccer balls that we had branded "The White House."

Then we boarded a flight to Kenya, during which Mrs. Trump had a specific request for us. While at the school in Malawi, the children had kept asking us to take photos of them on our phones so that they could see what they looked like. Suffice it to say, we were all surprised that many people there did not own mirrors and so the children literally didn't know what they looked like. As soon as we returned to the United States, she wanted us to send full-length mirrors to the school. "We need to send the school mirrors. Children need to know what they look like and see that they are very strong or very beautiful." She was insistent that the children should be able to look at themselves and know their self-worth. Sadly, that request was never fulfilled. According to her chief of staff at the time, the White House counsel felt that there was potential liability in sending mirrors. Only my opinion of course, but I never really believed that to be true, though, because on more than one occasion Lindsay shared with me that she thought it would be a PR nightmare, a model sending African children mirrors. Everyone in that White House thought they were an expert in communications; it was the story of my life.

AS MENTIONED EARLIER, THE Kenya visit was where her first solo interview as first lady—one year and eight months after she had taken on the role—would take place with Tom Llamas of ABC News. I had met Tom early on in the campaign days and had grown to respect both him and his producer, John Santucci, greatly. They were tough, but they were also fair. I thought her first interview should be with a male anchor, because I was hoping it would be more awkward for a man to ask a woman about her husband's alleged affairs, and we had agreed that no question would be off limits. We wanted to show the country that even with the Stormy Daniels issue, its first lady was fearless and open. The interview had been kept a closely guarded secret, as part of a negotiation, so ABC could properly promote it and we could try to contain any leaks or speculation. Our side agreed that on the day of the interview, ABC could start promoting it and then begin to use clips from the interview later that evening. I knew that our traveling press pool would be asking what she was doing that morning anyway. I also knew that they—and all of the networks—were about to be very pissed at me. The first interview with Melania Trump was a very big get, and we had had requests from everyone on a consistent basis.

That evening, I went to Mrs. Trump's room to do a little more interview prep. She rarely did speech prep, and as it was her first TV interview, we'd never done interview prep. For someone who liked to be prepared and was such a perfectionist, I think she hated the idea that she needed help with anything. I had given her a binder a week before we departed for Africa with all the questions I thought the interviewers might ask, along with potential answers, and we had gone over them on the plane a couple of times. Naturally I anticipated questions about the president's alleged indiscretions and of course on the jacket, her views on immigration, and the contrast between her husband's Twitter account and her online safety initiative. She was tired, but she did humor me for about an hour. I knew that the Stormy Daniels and Karen McDougal topics would come up, so I focused mainly on those and talked to her about when to pause and when not to. When she was nervous she tended to rush through things, and I wanted her to come across as the thoughtful, poised woman that she is.

I went to bed in my hotel room that night with the intention of getting up early and heading to the interview site before the rest of the team so I could look over the shots and lighting one last time and take some photos. Anytime she was going to be in front of the camera, even if just standing next to POTUS, I would send her pictures of the stage or setup so she could move things or have me adjust the lighting if necessary. Both she and the president were particular about lighting, always wanting it to be as "warm" as possible. I had been asleep maybe two hours when my phone started blowing up. That was never a good sign.

When I grabbed my phone, the first thing I saw was a text from the producer that said, "We need to talk," then an email from James Goldston, the president of ABC News, apologizing to me. Apparently there had been some kind of confusion with the time differences and ABC's PR department had sent out the press release announcing the interview twelve hours early. The rest of my texts and emails were from angry members of the traveling press pool and all of the networks, some of whom felt I had been dishonest and others who were angry that they hadn't received a heads-up. It was around 3:00 a.m. when I called the producer to talk through what had happened. I wasn't happy, and my gut told me that Mrs. Trump would be pissed. Our conversation was heated but professional, and I let him know that I'd talk to her when she woke up in a couple of hours, but I suspected she might not go through with the interview.

When I got to her door later that morning, I had a curler in my hair and was wearing a robe along with my Secret Service hard pin and slippers. Telling the agent at the door "We will never speak of what I look like," I walked into her suite to find her in the exact same outfit—no hard pin, of course, but including the curler on the top of her head. She didn't believe that ABC had issued the press release by accident, she told me. "I do not believe that was a mistake. They were afraid it would leak and released it. I do not like this." She felt we had given the network so much access on the trip and had agreed to any questions its reporter wanted to ask, and we were being taken advantage of. I told her mistakes do happen and I trusted those guys. She curtly let me know that she would still get ready and followed up with "I will tell you later if I want to do this still or we just go to the elephants," which was scheduled to happen right after the interview wrapped. My stomach dropped, and irate texts and emails from the press kept streaming in.

When I arrived at the interview site, I explained to Tom, his producer, and James Goldston where we were and that she hadn't decided if she wanted to do the interview. Goldston was apologetic and rightly worried, and the producer was pissed—at me, at his PR department, I don't know. After an hour of heated conversation, I was told that Mrs. Trump was on her way and we were good to go.

As soon as the first lady arrived, we went into her hold room to go over last-minute details. She wore tight khaki pants, high riding boots, and a crisp white shirt, and she was holding a white hat that looked exactly like what would appear if you googled "safari hat." Lindsay Reynolds joined us, as did Mrs. Trump's valet and hairstylist. She put the hat on, asking all of us "What do you think?," then listened as the four of us went back and forth while she looked into a mirror. "It will sit on the table next to me in interview, yes?" I liked the hat, but I worried that she might look as though she was in a costume, as though we had dressed her up as Safari Barbie. To be honest, as we had an entire TV news crew and set waiting on us, I wanted her focused on the interview and didn't want to say anything to break her confidence. One person in the group ended the discussion with a very definitive "Ma'am, they're going to beat you up no matter what. I say wear it. You look fabulous, and you're a fashion icon!," and with that she gave us one of her intoxicating laughs. Little did we know that a mere ten hours later, the hat would turn out to be our third major fashion blunder and would dominate the headlines as "racist" and "out of touch."

The interview went well, despite having to stop and tape some gaps in her button-down shirt closed a couple of times. After that interview we visited and fed baby elephants, which was incredibly fun for me and I think scary as hell for her. Remember, the Trumps are not really animal people—and baby elephants weigh upward of 250 pounds and get really rambunctious when it is feeding time. In fact, there was a moment when one of them jumped a bit, Mrs. Trump stumbled back a foot or two, and her lead agent caught her. The photo of that moment would later become a meme that our staff and the agent's friends loved to tease him with. Since he was standing behind her when she stumbled, he had to grab her around her chest, so you can imagine the subsequent jokes!

We took a safari tour before heading to the Nest, an orphanage in Nairobi. It is an incredible place that not only houses orphaned children but also provides facilities for mothers who have been incarcerated. They provide skills training for the women so they can provide for themselves and their children, then do all they can to reunite them. As I had been on the preadvance, I knew what a special place it was and was looking forward to seeing the babies again. Mrs. Trump glowed. She was kind, empathetic, and caring, asking questions like "How long do they stay?" "What do they eat?" "How are you funded?" and what the rate of reunification with mothers was. She eagerly showed the newborns as much physical affection as she could and was visibly impacted by some of their stories. One baby we saw had arrived only a week before after having been found abandoned in a sewage ditch. The older children danced and sang and presented FLOTUS with roses and smiles while we as staff looked on with tears in our eyes.

FROM KENYA WE HEADED to Egypt. As I had been on the preadvance, that was the portion of the trip I dreaded the most. The government was highly suspicious and not friendly, especially to women and a free press. Mrs. Trump met with Egyptian president Abdel Fattah el-Sisi and his wife, Entissar Amer, which went well; then we were off to the great pyramids. By that time, the "news" had hit about what Mrs. Trump had worn on the Kenyan safari, including the safari hat we had debated over.

Apparently the hat Mrs. Trump had worn on the safari was also known as a pith helmet, which to her critics meant she was deliberately brandishing a symbol of colonialism and white supremacy. In addition to that ridiculous story, I was still dealing with a traveling press pool that was pissed off that they hadn't been given more access to Mrs. Trump, something they weren't only making clear to me personally but many were using their favorite mode to complain: Twitter. I rode in the car with Mrs. Trump on the way to the pyramids. We didn't have a long drive, so as soon as she asked the now-common question "What's new?," I dived in. "The press pool is getting very irritated that they have not been able to speak to you yet, and they are still angry that you did the ABC interview, so you should probably stop on the way to the pyramids and talk to them on the record." She rolled her eyes, and I continued, "I know. But the news right now is all about the helmet you wore on the safari, and it is taking away from the great things you have been doing here. They are painting the pith helmet as racist." She responded, "That is not right. We were just at the slave castle in Ghana. Are they crazy to say I am racist?" Again I tried to be sympathetic while still making my point. "I know, but they are angry and we do need to give them something." "What do I need to tell them or talk about?" she asked, and I replied, "They will certainly ask you about the hat you wore on safari. I think you need to drive home the point that people should focus on all that you are doing for children and on this trip, not on your clothes. It's silly." I also let her know that they'd likely ask how she liked the trip so far and might bring up the alleged affairs (I always wanted her to be prepared for worst-case scenarios). She said, "I will take three questions only and tell them I wish they should focus on what I do, not what I wear." I agreed to her proposal and said that I would say, "Last question," so she'd know when it was time to walk on. Trying to wrap an interview for both the first lady and the president was always difficult. Neither of them liked the perception that the staff was "handling" them in any way, and the president often continued talking long after an interview was wrapped. Predictably, Mrs. Trump ignored my "last question" call and took a few more. As I have said, the two of them can be so similar, it is almost comical.

As soon as she got out of her car and with the Great Sphinx behind her, she took a couple of questions—and when the inevitable clothing inquiry came, I was ecstatic when she gave the answer she had practiced in the car verbatim (and had already given ABC). Unlike her husband, Mrs. Trump was excellent at staying succinct and on message. My happiness was quickly shattered when she proceeded to walk straight down to the Great Sphinx of Giza in her glamorous pantsuit, wearing yet another hat, and began to strike multiple poses. Everything she had just said to the press about focusing on her work over her clothes was being undermined by her standing in that outfit with the Great Sphinx looming over her, not saying a word to the dozens of members of the media who were gathered. Instead of talking about why we were there and all that was being done to conserve the Egyptian relics so that tourism could continue to thrive in Egypt or mentioning the work US-AID in our own country was doing to help, Mrs. Trump looked as though she were at a high-fashion photo shoot. It was one of the best and worst things about her from a communications perspective: she was disciplined and stayed on message when it was planned out, but she never did anything off the cuff, instead just standing for photos and ignoring shouted questions.

But overall, the trip went well: four countries in six days, impactful visits, much laughter, a few tears, and very little sleep. No one on the team walked away from it without having been deeply and positively affected in some way. I would love to devote the majority of this book to our time in Africa; the people, the places, the customs, the huge hearts and unassuming nature of the people I met there impacted me in ways I'd never thought possible. We could learn so much from them in terms of kindness and simplicity, and one day I do hope to live there and help the people prosper even further.

Of course, as we were on our very long flight back to the United States, Mrs. Trump noticed that the day after we had made our trip to the orphanage in Kenya and she had been photographed holding an African baby, Ivanka had been in North Carolina for a hurricane recovery visit, holding an African American child. Was that a coincidence? Probably. But it was an unwritten rule in the White House to try not to overlap events when the president or first lady was traveling, ensuring focus on just one visit. In fact, we had specifically asked the West Wing not to plan any large events while we were away. Yet there was Ivanka, once again doing what could be considered the duty of the first lady and posing for the media, while the real first lady was on her first solo international trip. That was the kind of dig we dealt with all the time. Of course, it was very noble for Ivanka to visit communities harmed by the hurricane. But Ivanka is smart, media savvy, and intelligent, and there was no way she didn't know our itinerary. Her visit could have waited—but when the Princess wanted to do something, she did it.

THAT NOVEMBER, SHORTLY AFTER the Africa trip, our biggest "newsmaker" statement out of the East Wing occurred when, seemingly out of the blue, to reporters at least, the first lady called for the firing of Mira Ricardel, a top adviser to John Bolton at the National Security Council. The announcement came in an email with my name on it that read, "It is the position of the Office of the First Lady that she [Ricardel] no longer deserves the honor of serving in this White House." Normally, of course, a first lady would never publicly interfere in the president's staff decisions. We knew it was an unusual action—CNN called it "unprecedented"; *Vanity Fair* called it "stunning" and "cruel"—but it was a truly unusual circumstance.

Ms. Ricardel was a stern-looking middle-aged blond bureaucrat perceived by some in the White House as someone who didn't navigate internal politics well or "play well with others." She butted heads with the likes of John Kelly and General Mattis, whom I heard she bad-mouthed and apparently disliked. She was, by many accounts, abrasive and sharp elbowed, but Bolton liked and protected her. That made sense. She was very "establishment," having worked for Senator Bob Dole in the 1990s and later in the Pentagon during the George W. Bush administration. I had no idea who she was until I was in her crosshairs.

One day I was informed that Mira was asking for an investigation into behavior on the preadvance trip to Africa prior to the first lady's first solo international trip. As I would come to learn later on, two things were often weaponized and used against people in the Trump administration, even if there was no reason: the security clearance process and the launch of internal investigations. Either could be deadly to a person's reputation, and the news of them often leaked to sympathetic reporters in the press. All anyone had to do to launch such an attack was mention inappropriate behavior to an allied senior staff person and an investigation was opened, which in turn could cost the person accused his or her security clearance or even job. The outcome of the investigation was often beside the point. Just being "under investigation" hurt your reputation inside and outside the building. Turned out it was the first of two times that the tactic would be used on me.

The story I'm about to tell would be funny and frankly lame if it hadn't had such potential long-term ramifications. The long and the short of it is that a group of us were sent to scout the first lady's trip to Africa, meaning tour various sites and then make decisions based on policy, logistics, and security prior to her arrival. Because the East Wing wasn't exactly policy driven and it was our first solo international trip, the NSC sent someone from its Africa desk to help navigate some of the policy issues and help with introductions at the embassies on the ground. Now, the first lady didn't really have any interest in what the NSC wanted her to say, but we tried to be polite.

The trip was grueling; we were in four countries in as many days, with twelve-to-thirteen-hour workdays and some nasty jet lag. Two of the nights we slept on the plane, and the various sites in each country were not only far apart and logistically challenging but emotionally difficult because of the circumstances in which we saw so many children living.

By the time the trip was done, we were exhausted, dirty, and ready to get home but would have to stop in Cyprus for the night because of flight crew rest requirements. We arrived at a decent hour in the late afternoon and agreed that after taking some time to shower and freshen up, we'd have dinner and drinks together. Roughly fifteen of us ended up at a beachside restaurant with delicious food and perfect weather.

After dinner, most of the group went to a bar for some drinks, where we promptly ran into a bachelorette party. The evening that ensued was actually so wholesome that it was laughable. We danced, started a conga line with the bride-to-be, and because she was so excited about an eighties song that came on, one woman on our team accidentally kicked her shoe onto someone's table. It was just a fun night—no one got overly intoxicated, no one got sick, no one embarrassed themselves, and we were all back at the hotel by midnight. The next morning, as we waited for our bus to the plane, we regaled the few people who hadn't accompanied us the night before with tales from the evening. Somehow the news made its way back to Mira Ricardel with predictable embellishments—that there had been drug use, public intoxication, and lewd behavior. I don't know for sure, but assumed the source was most likely one of Mira's aides, whom we had been forced to bring on the trip. She hadn't been out with us that evening but if it was her, she had decided she'd heard enough about it the next day to tattle on us and sex up the story.

Mira asked that the officious-sounding Office of Administration open an investigation into everyone who had been on the trip. Keep in mind that that included the first lady's staff, members of the Secret Service, members of the advance and travel offices, employees of the State Department, and her own NSC staff. Why it was under her purview or any of her business for that matter was beyond me. Needless to say, I was livid, as was our chief of staff, who said to either talk to everyone on the trip or shut it down. I volunteered to speak to anyone necessary, as did many others who had been on the trip, but it hung over our heads. We went to General Kelly, who ordered Bolton to have it stopped, which he ignored. Bolton was another hard-charging guy, a curmudgeon who didn't like being told what to do by anybody. We then went to the first lady, who also urged everyone to stand down and end the silly investigation.

Yet over and over, despite directives from both General Kelly and the first lady, the made-up investigation continued and, of course, news of it started leaking to the press.

I really don't know what the motive was. I assume part of it was her anger that one of her NSC people had not gotten a seat on Mrs. Trump's plane for the final trip to Africa. She probably also didn't like the fact that we had basically blown off the talking points the NSC had wanted Mrs. Trump to use. Or maybe, just maybe, she believed what she had apparently been told. But why it became a vendetta is a mystery. Maybe she didn't like being told what to do. Maybe she was just causing trouble. Maybe her inner Inspector Javert thought she was justified. But anyone in her right mind should know that you don't piss off the first lady. Mrs. Trump knew as well as anyone what something like this can do to people, especially staff members she cared about. As for me, I was already sensitive about the story because of my past DUIs, and I knew that people and the press would choose to believe the version of events with the most intrigue rather than boring facts, such as a fun night out with coworkers doing absolutely nothing wrong.

THE DAY AND THE way that Mira was transferred out of the White House was actually—as usual—totally unplanned. Mrs. Trump, her chief of staff, and I were having a meeting in the Map Room— one of our irregular sit-downs—and the topic of the "investiga- tion" came up. We explained where things stood and tried to think of ways to get the nonsense to stop. I was angry and frus- trated, and so was Mrs. Trump, who seemed mystified at this ran- dom woman's apparent vendetta. Also, I suspected the long-gone Stephanie Winston Wolkoff's repeated warnings to Mrs. Trump that the West Wing was out to get her surely played a role in the first lady's mind.

I have no idea how it popped into my head, but all of a sudden I said, "Look, I've been asked about this disagreement with Mira by the press. Why don't I just give someone a statement saying it is the position of the Office of the First Lady that she no longer deserves the honor of serving in this White House." I explained that it would make news instantly and once it was out there, Bolton would no longer be able to disrespect the chief of staff, and our office and the situation would have to be dealt with. Bold? Definitely. Unprofessional? Perhaps. But again, that White House played dirty, and sometimes you had to be willing to get your hands dirty right back if you wanted to defend yourself. Maybe that makes zero sense, but it was our reality.

Mrs. Trump thought about it for maybe two minutes then told me to do it. So right there in the Map Room, I sent the statement. Three minutes passed, and it was all over Twitter, and within ten it was on TV. Not more than fifteen minutes had passed when the outer Oval Office called and said that Lindsay Reynolds and I had been requested in the Oval Office.

Oh, shit.

Hearing the news that we were once again basically being called to the principal's office left the first lady almost amused. I will never forget her face as she looked at us and said nonchalantly, "Don't worry about it" and "Be strong!" as she let us walk over there to be yelled at by the president all by ourselves.

Right before we walked into the Oval Office, I got a text from her. Another "strong arm" emoji. Gee, thanks, ma'am.

As we entered the Oval Office, Trump was behind his desk on the phone, and I could instantly tell by the way he was talking that the first lady was on the other end. Maybe she had our back after all. But he looked annoyed. I could tell that he didn't like what she was saying, and he ended the conversation quickly as he saw us come in. We sat down across from him, and he threw a piece of paper at me; someone had printed my statement out for him. He was pissed and yelled, "What a stupid statement! Did you send this out?"

Obviously I had to tell him yes. Then, because, of course, he wanted things to be as awkward as possible, he called for Mira to be brought in. She arrived within minutes.

There the three of us sat. I was in the middle with Lindsay on my left, Mira on my right, and the president of the United States on the other side of the Resolute Desk. What ensued can only be described as a version of *The Apprentice: White House Edition*. Trump sat behind his desk, hearing each of us out as if we were contestants on the show and reserving his judgment to the end. As in the TV show, we had no idea what he was thinking or what the outcome might be. He could have said "You're fired!" to any of us.

At one point I remember quite clearly Mira pleading her case to the president, saying that in her entire seven months with him she had never leaked a thing.

To which I responded, "Well, I've been with you for almost three years sir, so if length of time is a competition, I win." Trump nodded as if I'd scored a point.

Mira then suggested that the statement be retracted by me with an apology. I let her know subtly that it hadn't been sent without permission and a retraction was not going to happen.

The president watched us argue some more, then yelled at all of us some more, and frankly, he was right to. There was no reason that the most powerful man in the world should have been bothered with such a stupid cat fight, though I do think part of him enjoyed it.

He ended up dismissing Lindsay and me after letting us know once again how stupid we had been. "Stupid statement," he said. "Really stupid. I'm tired of having to clean up these messes." This was another Trump tendency—to believe that he was some genius "Mr. Fix-it" with the media. As he had with Mrs. Trump's jacket, he thought himself the master at spinning the press and distracting them from one thing by throwing out some tweet or statement about something else.

But though he thought that what I had done was stupid, I didn't. It was bold—probably the boldest thing I'd ever done professionally—but the investigation was dishonest and wrong. Also, I had gotten what I wanted out of it. As we exited, I overheard the president tell Mira, "This isn't going to work. You're going to have to go." This really *was* an *Apprentice* episode.

I have no idea where Ms. Ricardel went after all that went down. I was told that Bolton kept her on the White House payroll for an additional six months without anyone else's knowledge after her assets were taken, and then I was told she turned down every job the White House offered her elsewhere in the administration, something that was done as a courtesy because Bolton and his top aide had never stopped pushing for it. Ambassador Bolton did not seem to like me after that and rarely acknowledged me. I believe he wrote a sentence about me in his book, calling me a "fury," which I promptly looked up. In short, "fury" has a few meanings: an avenging deity in Greek mythology who torments people and inflicts plagues, an avenging spirit, and a spiteful woman. It can also mean "extreme fierceness," which is what I'll choose to believe he meant.

AROUND THE TIME OF the Mira Ricardel circus, we received sobering news.

On November 30, 2018, I was traveling with the president and first lady in Argentina when Lindsay Reynolds knocked on my door. I checked my clock; it was around 2:00 a.m.

"President Bush has died," she told me, choked up. I would have to start getting into action with the White House response.

Our team in the East Wing had known for some time that President George H. W. Bush was in poor health. In part, that was because Lindsay had worked for the family and kept in touch with them. Although this sounds morbid, we had a statement prepared and ready to go once his passing was announced. It also fell to the East Wing to organize the White House's role in the funeral—helping to arrange travel for cabinet members and, most important, the use of Air Force One for the former president and his family.

Now, President Trump did not like George H. W. Bush, nor did he have much use for any of the Bushes since they didn't typically say nice things about him. When we told him that Bush had died and we would have to use Air Force One to transport the Bush family to Washington for a state funeral, he grudgingly went along.

But we all knew, or at least we feared, that Trump would re-fuse the use of his plane if we gave him certain pieces of informa-tion. The first thing we didn't tell him was that President Bush's service dog, Sully, would be on board. The president didn't hate dogs, but he didn't much like them, either. He thought they shed everywhere and would not have liked the idea of hairs all over his traveling office. So we never mentioned Sully, and thankfully he didn't ask. What he didn't know, the theory went, couldn't hurt him. Or us.

The other thing we didn't tell him early on was how Presi-dent Bush's body would be transported to Washington, namely, on Air Force One. We knew he wouldn't be okay with that, even for a brief trip. Dead bodies, death, sickness—those things really seemed to creep him out. Had we told him, we suspected he would have insisted on a thorough, costly, time-consuming deep clean of the whole plane. So we kept that to ourselves, too. If Trump was curious as to how Bush's coffin made it to Washington, he never asked, as far as I know. In fact, this book may be the first and only time anyone has told President Trump that a corpse and a dog flew inside Air Force One, at the same time, on his watch.

There was one other awkward moment involving the Bush funeral. The family didn't really want the Trumps there, a mes-sage that was sent to us subtly through intermediaries, specifi-cally Lindsay Reynolds. The president and first lady went anyway. Part of the reason was that they had already been embarrassed twice that year on similar occasions, the president having been essentially disinvited from the funerals of former first lady Bar-bara Bush and Senator John McCain. In addition it seemed to me that Lindsay, with whom I'd grown close and been through a lot with by that point, was beginning to develop a tendency to in-vent "problems" only she could solve. With Bush's funeral, it was a similar situation: she got to swoop in and say she had helped broker the deal to "allow" the Trumps to attend. To be fair, that was what the Trump White House did to the people who worked there. In an administration careening from crisis to crisis, the best way to look good was to be seen as putting out fires—even if you'd lit them yourself.

In McCain's case, the president hadn't even wanted to order the nation's flags to be lowered to half-staff. It was Ivanka who convinced him to do it, in the end, which was the right thing to do. But of course she took it too far when she and Jared then showed up to the senator's funeral, where they weren't exactly welcome. It seemed that once again the two of them just wanted to be seen and praised, which I think annoyed the president since it made his absence even more obvious. But as usual he just rolled his eyes at their antics. He knew there was no stopping them.

10

Chief Three

A good leader takes a little more of the blame,
a little less than his share of the credit.

—ARNOLD H. GLASGOW

Friday, December 14, 2018, the president announced his newest chief of staff, tweeting "I am pleased to announce that Mick Mulvaney, Director of the Office of Management and Budget will be named the Acting White House Chief of Staff, replacing General John Kelly who has served our country with distinction. Mick has done an outstanding job while in the Administration. I look forward to working with him in this new capacity as we continue to MAKE AMERICA GREAT AGAIN! John will be staying until the end of the year. He is a GREAT PATRIOT and I want to personally thank him for his service!" To help you keep track, not two years into the administration, we were on our third chief of staff, but this time he had "Acting" in front of his title. We were on a new attorney general, too, after Jeff Sessions had finally resigned that November. And less than a week after General Kelly resigned, Secretary of Defense Jim Mattis resigned with a scorching letter to the president that got a ton of attention. So now we'd have a second defense secretary in two years. But that was a whole other drama that I wasn't really in on.

Those of us in the know in the White House had seen the writing on the wall for General Kelly for some time. His effort to put together an organization with military-style efficiency was doomed to failure, though it failed faster than I had anticipated. The president didn't like structure any more than he liked someone else having power over whom he saw and talked to. I believe he eventually felt isolated and rebelled like a kid with an overly strict parent.

As the president got pissed off, Jared and Ivanka seized the advantage. Many of us had been enjoying Kelly's futile war against those two—denying some of their travel requests, keeping them from setting one-on-one meetings with the president to benefit themselves—but his efforts were just that: futile. As Kelly kept telling them "no" and tried to keep them from circumventing the rest of the staff, the leaks against Kelly and his team seemed to increase. Trump would read the stories (which magically managed to get set down right in front of him), and then lose more faith in his chief. Don't forget, that is how a lot of people would rid themselves of perceived enemies inside the Administration, they simply leaked things to the press—true or not—then made sure the president saw the stories.

I can't say I was brokenhearted by the change. General Kelly was disciplined and he was no bullshit, and I really respected those qualities, but he wasn't perfect. He let everyone know that he hated the job, which didn't exactly instill confidence in him. Maybe because he was a control freak, or maybe because Trump drives everyone to extremes, he oversaw some activities that were frankly shitty.

For example, one day and seemingly out of the blue, Lindsay Reynolds walked into my office and closed the door, which was odd. She sat down with a very serious look on her face, which was again odd, and let me know that one of my closest friends in the White House was being let go at that very moment. She said that there was an "issue with his security clearance" and that was all she could say. The man, who had been nothing but loyal to the administration and the president for more than three years, was walked out of the White House by a woman from Human Resources, followed by an armed Secret Service agent, and not given a second thought.

He was made to appear like some sort of a criminal, especially since no one would explain to him or, more shockingly, to the first lady, why one of her staff members had been bounced from the administration so abruptly. Mrs. Trump tried to find out what the offense was, but as far as I know, the White House Counsel's Office told her nothing.

I heard much later what had really happened. My friend was gay. Whoever determined security clearances had come across his Grindr account and decided that some of the stuff on it would be "personally embarrassing" to Mrs. Trump. Without asking her, as far as I know, they attempted to ruin the man's career.

To this day, I don't know if the decision went all the way up to General Kelly, although he was the type of guy who liked top-to-bottom control, so it was certainly possible. I do know that one of Kelly's deputies was well aware of what had happened and supported the decision, something that enraged me on many levels.

From that point on I was professional but no longer friendly with the Kelly crew. I just couldn't get over that level of bias. It was also my first really bitter taste of the transactional nature of the Trump White House. Despite that person's loyalty and incredible work ethic, no one had stood up to fight for him. I'm not sure the first lady did, either. When I went to Mrs. Trump after I'd learned the real reason for his dismissal, she acted angry at the news, but nothing happened. If she had really wanted a staff member back on her team, if she had been truly outraged, she could have done something. She didn't. That was frustrating, and it should have warned me of things to come. Mostly, I was offended by the amorality of it. If the person had truly been removed solely because he was gay and had a lively Grindr account, that was wrong. This was a White House filled with adulterers. I had a DUI, and they let me stay. It seemed so wrong to do this to a person and, even worse, not seem to care about it. I started to hear that voice inside me again: "This is not okay," but this time I just blamed the chief's office.

Not long after that, John McEntee was frog-marched very publicly out of the White House for the same nonreason—a "security clearance" issue. It was reported that he had racked up a fair amount of gambling debt, and in the eyes of the Kelly crew, that was something a foreign country could have used as blackmail. That one stunned me, because if McEntee, the president's long-serving body man, who was as close to him as anyone was, could be walked out, any of us could. It seemed as though some of the most loyal and well-intentioned people to the president were being picked off one at a time and again, no one was doing anything about it.

McEntee was another one whom I stood behind and vocally defended. Loyalty was supposed to count for something in our administration. It did not win me any friends in the Kelly days, but I didn't care.

The greatest irony, of course, was when I heard that Jared Kushner never got a security clearance because of all of his issues, financial interests, and so on. As far as I know, he only ever got the most basic clearance, yet he was in all of the most important meetings of the administration. He also reached out directly to many heads of state, such as Benjamin Netanyahu and the leadership of the Saudi government, something no staffer should ever do, let alone if they had no top security clearance. Jared seemed pissed about the clearance issue, and I think he blamed General Kelly. Though he denied it, I believe the president ended up overriding General Kelly so that Jared and Ivanka were granted the very basic security level of Top Secret, or TS. To my knowledge neither of them ever achieved Top Secret/Sensitive Compartmented Information, or TS/SCI. Suffice it to say that once Jared broke Kelly, he became the de facto chief of staff of the Trump White House for the remainder of our time there.

After all was said and done, I will say this about John Kelly: he dedicated his life to our country, and he and his wife lost their son in service to this country. I remember sharing Thanksgiving dinner with Mrs. Kelly on Air Force One and staying up all night over bourbon talking about their son. She spoke with such pride and also pain about his growing up, his time in the military, and the aftermath of his passing. This country owes that family—and so many others—such gratitude. I believe that General Kelly is an honorable man with perhaps some old-fashioned ideals and that he tried to do a good job in his role as White House chief of staff. Kelly would be the cause of one of my biggest regrets in the Trump administration, but more on that later.

I HAD NEVER MET Mick Mulvaney before he became chief of staff, or if I did, I don't remember it. He is of average height and weight, with light-colored hair and round glasses. Like me, he cannot remember anyone's name. He is a devout Catholic, loves golf and swimming, has a wicked sense of humor, and was always honest with us. He also has the most amazing wife, Pam. When Mick and the small staff he brought with him started on January 2, 2019, I was not initially impressed. I heard from some that the Mulvaney crewwere unprofessional, only wanted to have fun, demanded to see the bunker, and felt that Camp David was their playground—a far departure from General Kelly's style. But as I got to know the Mulvaney team over time, my mind began to change. Not only did every one of them bend over backward to be welcoming and relaxed, they were damn good at policy and staff issues, and their discretion was incredible.

They began by sending emails out on Friday, inviting everyone to stop by the chief's office for popcorn and soda. In a place that had been ruled with a military formality for more than a year, that was a breath of fresh air. Mick also had an open-door policy, and if he wasn't available, his principal deputy chief of staff, Emma Doyle, always was. At thirty years old Emma, by the way, was the youngest woman ever to have held that position, which is very impressive considering that she took Mick's place when he couldn't be around and was charged with continuity of government if some serious shit ever went down.

To me, the best part about Emma was that I never once questioned if she was up for that kind of responsibility. She was always calm, always composed, and pragmatic in the toughest of situations. I could see that the Mulvaney crew were coming into the White House with no predispositions. They genuinely wanted to know and understand everyone's roles and what the staff thought worked and what didn't.

II

The Princess and the Queen

None are so empty as those who are full of themselves.

—BENJAMIN WHICHCOTE

In early June 2019, President and Mrs. Trump were invited to a state dinner in the United Kingdom hosted by Queen Elizabeth II. All of us at the senior staff level would get to attend—as well as meet the queen, Prince Charles, Prince William, Kate, Prince Harry, and the rest of the royal family. We were all beyond excited, but none more than the Trump children. It was the only trip in all four years that every single member of the family expected to attend, including Javanka; Eric and Lara; Don Jr. and Kimberly Guilfoyle, his girlfriend and Fox News fixture; and Tiffany and her boyfriend. All of them wanted seats at the state dinner that the queen would be hosting, as well as the dinner at Winfield House, the US ambassador's residence, which would also be attended by Prince Charles and Camilla, the Duchess of Cornwall. Because of course, everyone began maneuvering almost immediately.

In the first place, the logistics were incredibly complicated. There was barely enough room on the trip for the presidential support staff. Adding the Trump kids and their spouses and their security details would push off people who needed to be there to do their jobs. There was also a limited number of hotel rooms and vehicles for motorcades. All in all, we were not set up for a smooth planning process. It set the stage for a disaster internally.

That trip is a good example—not that I needed another—of the complications caused by Jared and Ivanka's various ideas and requests. With seating and accommodations so tight, "the interns" used their role as family members to intimidate those in the advance and travel offices to allow them to have their own staff accompany them. It was a running joke in the East Wing that Ivanka and Jared each had a chief of staff. In a normal White House, only the two principals—the president and vice president—have a chief of staff. In the East Wing, the first and second ladies generally do, too. But in the Trump White House, having one's own chief of staff was a cool affectation.

Lindsay Reynolds had dreamed of meeting the queen since forever, and in her role, she was hell-bent on making sure the Trump kids didn't ruin it for her just so they could be "seen." In particular, it seemed that she intended to move heaven and earth to keep the kids off of Air Force One and Kimberly as well as Tiffany's boyfriend out of the official state dinner. "We are going to look like the Beverly Hillbillies," she told me. "We'll be an embarrassment to the whole country."

Lindsay did ultimately succeed in keeping them off Air Force One, but her victory had repercussions. Don Jr. and Eric made their displeasure known to their father and when Lindsay got on board Marine One with the first lady to head to Joint Base Andrews, the president blew up at her. He told her that his children are "flying on a commercial plane right now" (as if that was the greatest burden any person ever had to bear) and she better never "fuck with his kids again."

As the president launched his verbal barrage, Mrs. Trump tried to intervene. "Don't talk to her like that," she said. But the president didn't care; he was pissed. He let it be known that if anything else went wrong with his children on that trip, Lindsay would not be coming back to the United States with us. She was on his shit list for a long while after that and I felt terrible for her. I think we didn't realize that Trump was already sick of having his kids bitch to him about the trip and so our staff was a convenient outlet for his frustrations.

ANOTHER FIGHT THAT WAS simmering prior to the trip was Jared and Ivanka's expectation to join the president and first lady in the initial greeting with the queen. Protocol dictated that the queen greet the president and his wife alone as they arrived to meet her and then meet the other family members and staff later. Ivanka and Jared's apparent ask was completely inappropriate. To my knowledge they also never explained why they alone—not Eric and Lara or Don Jr. or Tiffany—deserved that special honor.

The first lady, our favorite rule follower, was not having it for a moment. She was adamant that protocol be followed with the queen. "It is inappropriate, it should be just the president and I," she told us very clearly. There was no way in hell that "the interns" would walk side by side with her and the president after they arrived at Buckingham Palace by helicopter—and she was right. It didn't dawn on me at first, but over the course of the next few days, I finally figured out what was going on: Jared and Ivanka thought they were the royal family of the United States—on the same level as William and Kate in the United Kingdom. We didn't call her "the Princess" for nothing, after all.

The first lady also knew that her husband was getting more than a little irritated with his kids calling him to complain or ask for one thing or another on the trip. She was smart enough to stand clear of that. So though it was never said out loud, I think she expected us to figure out a way to fix the problem without having to complain to him herself. "I will just be the bitch who hates his kids, and that is not true. I don't want to deal with this again," she said on that and many other occasions to follow.

In the end we were saved by the fact that there were not enough seats on the helicopter for Javanka, so they had to slum it with the rest of the staff. But I did watch with amusement as our delegation was waiting to go onto the balcony, when the two of them seemed to make sure to look out the windows often enough to be captured by the hordes of photographers. The pictures didn't turn out the way I think they were hoping. The images ended up making them seem pathetic as they stared out longingly, likely wishing they were the first couple.

Still, even with those sad sacks taking up all of the oxygen, the official welcome at Buckingham Palace was incredible. The senior staff delegation watched as the presidential helicopter landed on one of the expansive lawns at Buckingham Palace and the president and Mrs. Trump disembarked. Waiting to greet them were the queen, her husband, Prince Philip, Prince Charles, and Camilla.

Mrs. Trump was in a brilliant white suit with what NBC called a "stunning, and stylishly tilted, wide-brimmed hat." The dress was from one of her favorites, Dolce & Gabbana, but the hat was pure Hervé Pierre. It had traveled very carefully in a box from the United States.

After the arrival ceremony we were ushered into a very ornate room and lined up so that we could meet the queen. She was tiny, just as in every picture I'd seen of her. Her purse was hanging from her arm, and she was quite formal with her quick hello.

Prince Charles followed her, and I was struck by how engaging he was. Well briefed, he knew I had worked for both the president and first lady and asked me, in his cultured accent, "How do you find time to do anything at all?" The president liked the queen a lot, but he was apparently not as big a fan of Charles. On the trip he and the first lady went to meet with him for a private tea. After Trump returned, he complained that the conversation had been terrible. "Nothing but climate change," he groused, rolling his eyes. Mrs. Trump laughed and said of her husband, "Oh, yes, he was very bored." She, on the other hand, had a lovely time.

THE STATE BANQUET THAT evening was a white-tie formal affair. There is something magical about arriving at Buckingham Palace dressed in your very best. As with everything else having to do with the monarchy, there were a ritual and a cadence to the events. Every move was choreographed, and that was apparent when our delegation was put into a line to be paired with our tablemates so that we could be announced as we walked into the room. It was intimidating as hell because all of the guests who had entered ahead of you were already seated, and as you walked into the room, all eyes were on you. I'd like to believe I looked stunning and elegant, but truth be told, I was just trying not to trip on my gown. I was wearing an off-the-shoulder gold sequined dress with a bat sleeve on the other arm. I'd had it altered and tailored to fit by a seamstress at the White House but had changed shoes at the last minute, so it was a bit long. It was an off-the-rack dress I'd seen at Macy's one day and immediately fallen in love with.

The room where the state dinner took place was huge, with the table arranged like a horseshoe. There were roughly 170 guests in attendance, and the table was decorated with lavish bouquets that included peonies, larkspur, sweet peas, and roses. It was reported that the menu had taken six months to plan and the table three days to set. An orchestra played on the balcony above us, and though I cannot remember the titles, there were definitely a few show tunes thrown in there, no doubt for the president.

The queen and Prince Philip sat at the head of the table, flanked by President and Mrs. Trump. There was some gossip later in the press about the queen's choice of headgear for the evening, the Burmese ruby tiara. *Vogue* pointed out that the exquisite piece, according to the royal curators, was seen by the Burmese as having "prophylactic properties guarding the wearer not only against illness but also against evil." That was interpreted by the press, naturally, as Her Majesty trying to ward off Trump's evil.

Aside from the youngest, of course, all of the Trump children and spouses had made it to the dinner, minus Kimberly Guilfoyle and Tiffany's boyfriend. To my left was a gentleman who worked at the equivalent of our USDA and to my right was Princess Michael of Kent. I don't know how else to say it other than I lucked out on my tablemate. Usually at these kinds of formal affairs, the conversation is stale and awkward—but not with Princess Michael. At seventy-four years of age, she was loud and lovely, eccentric but with a great sense of humor. She delighted in telling me repeatedly that she had never changed any of her children's diapers, and she gossiped to me about Princess Diana, saying how vulnerable she had been and that she had been miserable in the marriage right from the start. She also told me that the royal family had never been welcoming to Diana, who had basically been there to provide an heir to the throne. As she was married to the queen's first cousin, I kept wondering "Should she be telling me this?" She also spoke to me at length about her book, *A Cheetah's Tale*, and we spoke about Africa before she turned back to the "tired nature of the monarchy." At the end of the dinner, cherries were served as dessert, and I marveled at how elegantly the woman was able to take the pit out of her mouth and place it in a bowl.

I would discover at the after-dinner reception that Princess Michael was perhaps something of a black sheep in the family. Every time I was asked whom I had sat next to—including by Catherine, duchess of Cambridge, otherwise known as Kate Middleton—I was met with an odd look and a tentative "She is very interesting, isn't she?"

Prince William and Kate both had a lovely sense of humor. I was wearing a headband that mimicked a tiara because I wanted to pay homage to their way of dress, something I had learned from Mrs. Trump. Kate complimented me on it, and I thanked her, saying "It's certainly not as beautiful as the one you're wearing," to which she replied, "Yes, but it's also not as heavy as this." William chimed in with "I'm glad I'm not required to wear those things—though I'm not sure I have the hair required to keep it on." I was struck by how thin Kate was; the two of them seemed genuinely in love, or at least they put on a very good show of it. Prince Harry was also there, sans Meghan Markle, which surprised no one as she had made it very clear that she did not care for our president. He and I had met before, when Mrs. Trump had traveled to the Invictus Games, and we had a great conversation about that event. He joked with me about the "stuffiness" of the evening, which seemed to be a theme with everyone there. Of all the royals he was my favorite because his sense of humor and down-to-earth style put me at ease.

THE NEXT NIGHT WAS to be a dinner at the US ambassador's residence, Winfield House, set on a massive twelve-acre plot in the middle of Regent's Park. Technically, the dinner would be hosted by the president and the first lady, and both Prince Charles and Camilla would be in attendance. The evening started off with a good bit of drama. The security assets were stretched thin because of all the protectees who were traveling with us: the Trump kids, Ambassador John Bolton, Chief of Staff Mick Mulvaney, and Secretary of the Treasury Steven Mnuchin. It was arranged that the entire delegation would be put onto a bus and Ambassador Bolton's car would lead the way with lights and sirens to ensure that we would avoid traffic and get there on time. We did not want to keep members of the royal family waiting.

But when the time came, Bolton, notorious for being difficult and "not playing well with others," didn't show up. Nor did he answer his phone, and his Secret Service detail wasn't sure what he was doing. My guess was that, hearing of the plan, he didn't want to be the head of the "lesser staff" parade. Because we had no escort, we left on the bus about ten minutes early and slowly made our way through London in black tie to get to Winfield House. When we were a little more than halfway there, Ambassador Bolton's car zoomed past us, lights and sirens blaring. I will never forget the anger on Chief of Staff Mulvaney's face as he watched Bolton's SUV drive past us and said, "That fucking prick." When we finally arrived at Winfield House, Mulvaney sought out Bolton, and the two of them got into a heated argument over his antics. I couldn't quite hear what was said, but I know that we had to close doors so no one else could, either.

Right before the dinner started, we were once again put into a line to greet Prince Charles and Camilla. When it was my turn to shake hands with Charles, I was delighted that he remembered me from day one, saying "Here you are again—have you managed any rest at all?" He was lovely and very personable, something that doesn't come across when you see him on TV. At dinner, I sat next to Sir Nigel Kim Darroch, who served as the British ambassador to the United States. We bonded over wine and American football, but I did get an odd vibe from him when he asked, "How do you do it? Work for a man like your president?" I guess it was no surprise when he ended up resigning the next month after confidential emails of his saying that President Trump "radiated insecurity" and describing the White House as "inept, dysfunctional, chaotic" were leaked to the press and became very public.

At the annual White House Easter Egg Roll. I knew how sensitive the Trumps were to sickness or death, so I awkwardly diverted their attention from a man only a few feet away who had suffered a heart attack by literally telling them to "look over *here*." I was pointing at nothing. WHITE HOUSE PHOTO

Left: That damn jacket! On our way to the principal's office—I mean, the Oval—to get yelled at by POTUS. "What the hell were you thinking?!" he snapped as the jacket scandal consumed the media. Trump was almost marveling at how anyone could fuck something up as badly as we just had. WHITE HOUSE PHOTO

Right: I took this picture because I so rarely saw any public display of affection between the president and the first lady. Here she is leaning into him just before they go out to see a crowd at the US embassy in Paris. Later that day the president gave his wife a kiss on the lips—another rarity. (She was more of a European kiss-on-the-cheek person.)
COURTESY OF THE AUTHOR

Walking with the president right before he officially hired me as White House press secretary and communications director. I noted I'd still be working with the first lady, too. "But remember, I'm the only one who matters," he told me. Truer words were never spoken in the Trump White House.
WHITE HOUSE PHOTO

Getting ready for a speech in India, in the company of top Trump aides Dan Scavino (*second from right*) and Stephen Miller (*far right*). Both of them were masters at handling Trump's moods, especially Scavino. Whenever the president was losing it on someone, Scavino would chime in with good news, like, "Sir, you're up two points in the Rasmussen poll!"
WHITE HOUSE PHOTO

With White House social secretary Ricki Niceta (*right*) and the first lady (*left*), getting ready for the first state dinner with France. Trump wasn't big on the trim Macron, whom he once dismissed as "a wuss—all one hundred and twenty-five pounds of fury."

The president loved to tweet something and then see how quickly it ended up all over the media. "Watch this, kids," he would say. Here we are laughing at one of the million times this happened.

Going over a speech with the first lady. As I got to know her better, she seemed more comfortable asking me to help her with pronunciations so she wouldn't be laughed at. It was a rare and endearing expression of vulnerability from her. I wished more people got a glimpse of that.

I often stood right outside the president's private bathroom between the Oval Office and his private dining room while he washed his hands or put on makeup so I could brief him on one thing or another. I actually first met him in a bathroom, which is not as creepy as it might sound.
WHITE HOUSE PHOTO

The president liked to edit his own remarks and tweets. I would often be there and offer suggested edits, but he rejected them and did the opposite about 90 percent of the time.
WHITE HOUSE PHOTO

I snapped this photo of the first lady as we were approaching the pyramids in Egypt. I loved that she was taking her photos openly, just like a regular tourist.

At the pyramids, I told Mrs. Trump that she really needed to gaggle with the press. She said she wished people would talk about what she did rather than what she wore, although right after that she stood in front of the Great Sphinx and started posing like a model. Old habits die hard.

One of many walks the first lady and I shared on the colonnade between the East and West Wings.

WHITE HOUSE PHOTO

After the first lady gave her convention speech in the Rose Garden—the one I worked so hard on. The president told me the speech was so good that he was going to take me back to the West Wing. Mrs. Trump and I quickly shut that idea down. The first lady aside, I had grown thoroughly disillusioned with the president and everyone who worked for him and wanted to get the hell out of the White House.

WHITE HOUSE PHOTO

One of my very first duties as deputy press secretary on a flight to Mar-a-Lago. The president went back to speak with the press and made the first lady join, even though she didn't care to. It was one of my first interactions with her—she was very kind and smelled incredible.

WHITE HOUSE PHOTO

At the Wailing Wall with Ivanka (*second from right*)—of course Ivanka had to be there. I wrote the blessing that the first lady tucked into the wall. Our office was disappointed that the Kushners decided to attend with the first couple when they could have done it on their own, in private. WHITE HOUSE PHOTO

Walking down the colonnade with the president of the United States. By the end of my tenure as press secretary, I never knew what version of the president I would get from one minute to the next, so I was always on edge. WHITE HOUSE PHOTO

Speaking to the president with other members of senior staff right before he announced that terrorist Abu Bakr al-Baghdadi had been killed.
WHITE HOUSE PHOTO

Getting ready for a PSA about hurricane preparedness. I was helping the first lady with her hair—she always had to have it perfect, but we had fun while doing it. We definitely laughed a lot, but at the very end I felt let down by her.
WHITE HOUSE PHOTO

Three Jobs

In June, Sarah Huckabee Sanders announced that she was resigning as the president's press secretary. I was surprised by that. I had been hearing gossip that the president wasn't happy with her because of the ABC interview he had done with George Stephanopoulos, but such rumors were a dime a dozen, and I hadn't put much stock in them. That was the interview that had aired an outtake of the president getting annoyed at Mick Mulvaney coughing in the background.

The truth was that pretty much everyone eventually wore out their welcome with the president. We were bottles of milk with expiration dates. He either got bored with the person or got tired of hearing them tell him things he doesn't want to hear or decided to blame them for whatever problem he was having at the moment. Maybe Sarah decided to get out while he could still tolerate her. If so, she was a smart woman.

I had gotten to know both Sarah and her husband, Bryan, very well over the years, and I knew she had aspirations to run for governor in her home state of Arkansas. I also knew that her job had understandably taken a toll on her. Contrary to what most everyone believes, the job of press secretary isn't just giving briefings on TV. You are on call 24/7 to the president of the United States as well as by the press corps, which is not a small group of people who keep it to one question per day. You need to know what is happening in the news at all times, and you are constantly on the phone with heads of agencies, trying to figure stories out. In addition, the press secretary oversees a team of around ten people and is expected to attend most every meeting in the building. It is a time-intensive job, and with three small children, Sarah felt it was time to move on. Not to mention the very sad fact that she was the first press secretary in history to require Secret Service protection after the agency ascertained that there were enough credible threats against her and her family to warrant protection.

ALMOST IMMEDIATELY, NAMES BEGAN to circulate as to who would replace her. Hogan Gidley, Sarah's principal deputy, was the obvious next choice, and Jared and Ivanka were supportive of him. Hogan had been savvy about cultivating them the entire time he was in the White House, always helping shape positive stories on their behalf. More important, though, he had worked hard to quell or defend any negative press about them, which, as noted, was something the two of them not only appreciated but expected. Other names that were mentioned for the job included those of Heather Nauert, a former spokesperson at the State Department, and Tony Sayegh, who worked in communications at the Treasury Department.

I won't lie, I was thrilled when my name first popped up in the media, as my dream was always to be White House press secretary. And once it did pop up, I got excited about the opportunity and discussed it with people who were supportive of me. I'd always hung a picture of the White House in the offices of previous jobs I'd held so that I could stay focused on where I ultimately wanted to be. For a communications person, especially if you actually like politics, White House press secretary is the pinnacle of your profession. In any other administration, if you reach that milestone, you were pretty much guaranteed a lucrative job at a large company, in addition to TV contributor contracts, and speaking engagements once the administration went out. But the Trump White House was different—which may be the biggest understatement ever written.

Though the prospect of reaching my professional dream was beyond exciting to me, there were a number of issues that gave me pause about the job or at least diminished my chances of getting it. First, the president was his own spokesperson, and he always thought he could do the job better than anyone else. Second, I wasn't exactly close with Jared and Ivanka, though I don't think they knew how much I resented them at that point. Third, I loved working for Mrs. Trump. Finally, Hogan was a dear friend, so I was pulling for him.

As the days went by, my name surfaced as a candidate in the press more often. It got to the point that one day Sarah and Hogan called me into her office to ask if I was pursuing the job. I let them know that if it was offered to me, I would have to seriously consider it. Hogan said he obviously would be honored to have the job, but if it wasn't him, he'd want it to be me—and I said I felt the same. Sarah let us know that she was supportive of both of us. Looking back, that was probably the most adult and professional meeting I ever attended in my four years at the White House.

When Emma Doyle, Mick Mulvaney, and I began talking seriously about my taking over, I was up front that I would still want to work for Mrs. Trump. Not only did I like the job and had grown comfortable in the role, I had a stupid-ass idea in my head that I'd be able to change the president's mind on things or do things differently because he would know I had a direct line to his wife.

The other issue that needed to be tackled was the structure of the press and communications departments. Traditionally, the director of communications and the press secretary oversaw their own separate teams but worked together in tandem to ensure cohesive messaging. The Trump White House had started off with a different way of doing things: Hope Hicks was the director of strategic communications, and although she was part of all the press and comms meetings, she worked independently and closely with the president in general on larger interviews, or she would let the press secretary know what he was thinking or was on his mind on any given day. Mercedes Schlapp had joined the team but I heard she had requested the title of senior adviser for strategic communications, which was a smart move because apparently "senior adviser" titles put you higher in the order of precedence, meaning that at events and in delegations, Mercedes was considered more senior. Sarah and Mercy had very different work styles, Sarah being more of a workhorse, whereas Mercy favored meetings throughout the day, which seemed to end in changing messaging at the last minute.

My thought process was that I could streamline messaging and coverage decisions if I oversaw both teams and put really strong deputies on each side. I also felt it was important that after Sarah left, Mercedes understood that there was one person in charge. That was not at all a slight of Mercy, whom I got along with well, but it looked to me that the current structure hadn't been working for quite some time, and I really wanted to do things differently. Six months prior, the president had stopped all press briefings, so in my mind, I was going to get as many people as I could (myself included) on TV and focus on the regional communications teams so we could give local reporters across the country more access. That is, if I ended up with the job.

IT SURPRISED NO ONE that the president took his time making his decision. He relished the speculation and competition among staffers. He also enjoyed hearing the different points of view about each candidate, and frankly all of the gossip about them. I had already spoken to Mrs. Trump about the concept of my having three senior roles, including staying on as her communications person. I also told her if she didn't like the idea, I would happily stay with her—and I meant it. But she was supportive, as I think she, too, liked the idea of having West Wing resources and information available to her side of the house. I also think she was toying with the idea of getting a new communications person to focus on TV, so this could have been her gentle way of telling me to go.

The president finally offered me the job in a (characteristically) random way. He was walking out on the South Lawn, where a newly wrapped campaign tour bus with the Trump-Pence logo was parked. The campaign and White House advance office had recently had it completed so that the president and vice president, if they should choose, could hit the road on policy tours across the country. The White House was involved for security purposes even though the campaign had paid for it. To my knowledge the bus was never used. I have no idea what it cost.

I happened to be in the Cross Hall when the president walked in with his lead agent Tony Ornato and Jordan Karem, the president's body man. They invited me to come take a look at the bus, and then the boss and I walked together over the South Lawn toward the Oval Office. As we were walking, he brought up the communications jobs. "Do you really think you can do it?" he asked. He went on to deliver his typical spiel—how many "big names" were begging for the job and what a big deal it would be to get it.

"I'd be honored, sir" was my response, because "I'd be honored, but I am also scared shitless" wasn't really an option.

He continued, "You know you will be a star," which was what he typically said to people who went on TV and fought for him. I reminded him that my only goal was to serve the first couple, saying "In my mind there are only two stars in the White House—you and Mrs. Trump."

The president sometimes liked it when I referred to his wife as the boss—I think he found it amusing in a way. But as we reached the top of the steps to the Oval Office, he quickly said, "But it has to be me first now. Always me." Then he yelled for Shea Craighead, his official photographer, who was always a few steps away. "Shea, let's get a picture. This is a big day, an important day for Stephanie. Get a good one, okay, ready, quick, quick."

The picture is . . . not my best. I wasn't prepared for any of what was happening, and it was a hot, muggy day, and I fear I was very sweaty. My lack of confidence in my appearance is becoming a running theme of this book, I realize.

Awkward photo session complete, we walked into the Oval Office and sat down. He looked at me very seriously. "You know everyone likes you," he said. "I was frankly surprised by that."

I laughed.

He continued, "No, seriously, everyone I have spoken to about you likes you. There's usually always one or two people who have something bad to say about a person for a job like this, but not you. Everyone just loves you."

I wasn't completely sure that was true, but it was nice of him to say so. So I thanked him profusely and said I'd try not to let any of them down. Then he uttered something that was pure Donald Trump. "They don't matter," he said. "Only me."

I laughed awkwardly, but he wasn't kidding.

"And Mrs. Trump," I reminded him. He smiled in agreement.

We then called Mrs. Trump on speakerphone. "Hi, sweetheart," the president began, "I've got Stephanie here. She thinks she can be press secretary. It's a big job, it's a tough job. What do you think?"

Mrs. Trump said, "I know she will do a great job. When will you announce it?"

I jumped in: "What if you announce it on Twitter, ma'am? It will get a ton of attention and make it clear that you have blessed this. I also think it will look great that the East and West wings will be potentially working together a bit more." I knew the last part wasn't true, by the way, but it sounded good.

My comms mind also knew that it would show unity between them as a married couple, as well as between the two sides of the house. POTUS could then retweet her with a message of his own. Mrs. Trump loved the idea, and the president didn't seem to care one way or the other. The rest of it happened in a blur: the tweet went out, the media started reporting it, and the obligatory "congratulations" texts and emails started to roll in. I think the text that was most relatable came from a good friend of mine, Matt Cooper, who simply sent "Oh for fuck's sake" to a group thread. It made me laugh and also drove home the fact that I was really in it now.

I OFFICIALLY ASSUMED MY duties as communications director and press secretary to the president in June 2019. Sarah and I had one day of overlap before we left for South Korea, which was nowhere near enough to just try and understand the West Wing personalities. And while I had interacted extensively with the media in the East Wing, it was an entirely new ball game in my new roles.

The media. Where do I begin? It is not my intention to use this book to trash the press, nor is it to defend some of our administration's behaviors and actions against them. Truth be told, I think both parties were at fault for many things—including putting out misinformation. I have butted heads with members of the media for most of my career; that is part of my job description. However, I respect the mission that the press have, which is to give people honest and accurate information with no bias. I repeat, no bias. My ex-husband is a news anchor, so I am familiar with the news industry and have many friends who work as reporters. When I took the jobs of press secretary and director of communications, I knew what I was getting into. I had seen what both Sean and Sarah had dealt with, and I myself had dealt with some hostile reporters in my role with the first lady. Perhaps stupidly, I thought I would be able to take what I had seen and learned so far and do things differently. I thought that since public briefings had stopped, I would open the lines of communication by way of one-on-one meetings and offering up to the media subject-matter experts who could do deep dives on issues such as national security, the economy, and immigration. I truly believed the president was his own best spokesperson. He gave the press so much access and so many sound bites on an almost daily basis that I thought I would be able to focus more on print media, local or regional media, and empower my teams to support people throughout the administration to get some of the overlooked success stories out there.

I started things out strong, both literally and figuratively, on our trip to South Korea at the end of June, followed by an impromptu stop in North Korea. That was, in typical Trump White House fashion, an event that was last minute, random, and marked by misinformation. Before the trip to South Korea, Trump told reporters that we wouldn't be meeting with Kim Jong-un. Then, all of a sudden and after an offer to meet via Twitter, we were.

But back to the first stop, during the first press conference, the South Koreans wanted each spokesperson to choose which reporters would ask their respective leaders questions. When it came my turn to choose, I said, "I think I will let my president make his own choice." I was standing to the side when I said it, and I remember both leaders looking at me—one with surprise and the other with delight.

President Trump flashed a big smile, then said, "She knows me very well" and proceeded to take questions. When I saw Jared give me a wink of approval from the audience, I knew I'd done the right thing.

From South Korea, we headed to North Korea. As noted, the meeting had been arranged after the president had tweeted about it, creating complicated logistical and security issues for our operations team and the Secret Service and real consternation within the National Security Council. John Bolton apparently did not feel that the trip was appropriate, and I remember my shock at seeing him stay behind in protest as the motorcade rolled away from the South Korean palace.

Our visit to North Korea was historic. Not only did President Trump walk across the line into North Korea with Kim Jong-un, causing his Secret Service agents many heart attacks, it was also the time I physically scuffled with a North Korean security guard to let the press into the one-on-one meeting between the two leaders.

The moment we arrived at the demilitarized zone, more commonly known as the DMZ, things were chaotic. Three members of our advance team had gone ahead of us to negotiate everything from the length of time for the meetings to what chairs the leaders would sit in. Apparently for the North Koreans it was important that their "great leader" have a chair that would make him look tall. We arrived at the South Korean side and walked into a fairly nondescript building with a large open area and floor-to-ceiling glass windows and doors looking out to the North Korean side. Between the two sides, our two press pools had already assembled, the only problem being that of the small group of North Korean press, only one was a "legit" member of the media, though I say that lightly because it is state-run media. The rest of the North Korean "reporters" were there solely to grab intel on both the South Koreans and us.

The next thirty minutes consisted of our advance team, and at times me, chasing the pretend reporters around to get them to stay where they were supposed to. There was one petite woman who was especially quick, darting around with a ladder and her camera, taking pictures of everything and refusing to listen or make eye contact when we tried to corral her. The North Korean security guards were the same way: they would not make eye contact, nor would they listen to any kind of direction from anyone but their dear leader.

After the initial meeting and infamous walk to the North Korean side, the president and Kim made their way to the South Korean side to move inside for a short bilateral meeting. It was supposed to be closed to the press, but of course as they were walking over, President Trump invited everyone along. I followed Trump and Kim into the meeting room, and after introductions and photos were done with the dictator and Javanka (of course), the president asked Kim if it was okay to bring the press in. He nodded ever so slightly, and that was my cue. I went outside to find that the North Korean guards had already made a human wall of sorts, separating the press from the room where the two leaders sat. I immediately let our operations and Secret Service teams know that the media were to come in, then waved our press pool over.

All I know is that a bunch of bodies converged and our advance team was fighting off North Korean guards. I saw a small hole in the crowd and yelled for our press to hurry over. At that moment one of the guards grabbed me, throwing me to the side. I managed to wiggle away and elbow the man out just enough to get our pool into that room. Nothing hurt until it was all over and we all started comparing "injuries," which were really just marks on the skin from our giant tussle. Needless to say, I was hailed as a hero of press freedom that day. It was a wonderful way to be introduced to the world as the president's press secretary, but I immediately said to anyone who asked that the goodwill would likely last only a couple of days. That turned out to be no bullshit.

DURING THAT TRIP, I also came to know the president much better. We'd had a number of interactions before, as you've read, but all from the distance of the East Wing, where I wasn't in his 24/7 line of sight. That was going to change. As I knew from having watched Sean Spicer and Sarah Sanders, there was no job in the administration that came under as much direct scrutiny by Trump as that of press secretary and director of communications. And his press coverage was all he seemed to care about most days. (It was usually terrible.) The job also, by its nature, would give me a lot of face time with him.

My early impressions were positive. Trump could be funny and charming when he wanted to be, and everyone in a new job got a honeymoon period with the boss. Being a (relatively) normal person with close access to the most powerful person on the planet, I did what I felt anyone in my position would do: I repeatedly begged him to tell me if UFOs were real. To his credit, he never answered that question. But every once in a while, he'd give me a slight smile, as if he knew something, and then say, "I bet you'd like to know, right?"

I also observed his weird quirks (everyone already knows about the constant stream of Diet Cokes, of course). On our way back from the North Korean meeting, I joined him in his cabin on the plane and sat down in a chair right by his desk. We were in the middle of a normal conversation, but I could tell he was distracted by something. He kept looking at my chair or my hair, I could not tell which. Then I realized that one of his ties was draped over the back of the chair behind me. They were almost always one color, very long, and from his own Donald Trump tie collection brand that he had created a long time before.

"Do me a favor," he finally said. "I know your hair isn't greasy, but will you move that tie?"

Donald Trump loved his ties, and he was worried that whatever was in my hair might inadvertently touch it. From then on, I noticed he was obsessive about not getting anything onto his ties ever—and could be weirdly awkward about it.

He sometimes asked me strange questions. Once we were leaving the South Lawn on Marine One headed to Joint Base Andrews. The president and FLOTUS were in their usual seats, and Mick Mulvaney and I were on the small couch across from them. We were having a conversation about something, most likely media coverage, when all of a sudden the president stopped. He looked at me intently, with great seriousness, and asked, "Are your teeth real?" I was used to his getting bored or distracted in the middle of conversations or briefings, but that one caught me off guard. I told him that yes, they were real and I had never even worn braces. He seemed impressed by the news and also skeptical.

He went on to ask me how I had blue eyes and dark hair and asked where my parents were from. I naively responded, "Well, my mom is from Nebraska and my dad Colorado," and I remember his look of confusion or perhaps annoyance before he said, "No, like your ancestry, what is it? Because you have all the dark features with those eyes." I told him I thought we had French and maybe some German in our ancestry, but I really didn't know, which seemed to disappoint him.

On another Marine One excursion, I was in the same seat with the same people but minus Mulvaney. The president mentioned that I had a nice tan, and after I thanked him and explained that it was a spray tan, he asked, as if it was something he'd been thinking about for some time, "Are pantyhose something that older women wear? You don't wear them?"

Mrs. Trump and I laughed, and I said, "Yes, sir, my grandmother wore pantyhose. I don't think those are worn much anymore unless you're much older."

He then went on to talk about how beautiful Mrs. Trump's legs were (very true) and said, "Our beautiful and elegant first lady, better than even Jackie Kennedy, don't you agree?"

To which I replied, "Yes, sir."

On still another occasion, he asked me to reach out to a prominent Trump supporter in Arizona. He wanted me to advise her to no longer wear sleeveless dresses and tops, saying they weren't flattering to her and it wasn't "a good look."

"You talk to her, though," he said. "I can't with MeToo and all."

AS I MENTIONED EARLIER, Mrs. Trump valued her sleep on long plane rides. But the president rarely slept at all. That meant someone had to always be awake to keep him company, usually Dan Scavino, even if we were dog tired and jet-lagged. To spare Dan, we would take turns sitting with Trump and let him ramble on about whatever he wanted to. And when he was in a bad mood or a scary rage, we'd send for a staffer he referred to as "the Music Man."

"The Music Man," which was how Trump referred to him early on when he didn't know his real name, was in charge of putting together Trump's rally playlists. They contained the president's favorite songs—"Memory" from *Cats*, the Rolling Stones' "You Can't Always Get What You Want," and others—and the staffer would go in and play the music to distract or calm him down. It was sort of like soothing the savage beast.

"Go get the Music Man," Trump would say when he got bored with whatever or whomever he was dealing with, or felt a rage coming on, or he just needed to chill. And someone would fetch him—quickly.

Trump was also obsessed with the song "It's a Man's World" by James Brown. I talked him out of including the song in his rallies because it wouldn't have been a good look for him. So instead he would make us watch a black-and-white video on repeat that showed James Brown singing the song live. The president would provide a running commentary: "Look at that," "That's a great performer," "You know, they say he was abusive, I don't know." He would bob up and down to the song, and it seemed to put him in his Zen place.

Once when we were listening to music, the president, in kind of a zany mood, stood up, grabbed a speaker, and walked it over to the door to the bedroom in his cabin. The room was dark, since Mrs. Trump was sleeping (of course). Trump took the speaker, turned up "It's a Man's World" to the highest volume, and raised it over his head so that the music filled the bedroom.

Although I was laughing, part of me thought, "Oh, shit, I do not want to be here if Mrs. Trump gets up." She never did. He did it to be playful with his wife, not cruel, and she certainly didn't take it otherwise.

At another point she suggested that he add the song "Purple Rain" by Prince to his playlist, which he did. At a rally later, when "Purple Rain" came on, he turned to his body man. "Get me Melania, right away."

The guy got through to the first lady and handed his phone to the president. Trump held the phone out and said, "Darling, listen. 'Purple Rain.' "

"That's nice, Donald," I assume she replied.

He smiled and hung up. Then, instead of handing the guy's phone back to him, he dropped it on the ground without a word and walked off.

He had that side to him that I might even call sweet. At least that was how I wanted to see it then.

Those first few days and weeks were what I had always hoped I would be able to do as a White House press secretary: work closely with the nation's president and have a front-row seat to watch him make history. I couldn't imagine that I could go anywhere but down after that, and in that I wasn't disappointed.

13

The West Wing

He intimidates people because he will attack viciously and
relentlessly. Yet somehow people crave his approval.

—MIKE DUHAIME

Our return from the North Korea trip marked when I rolled my sleeves up and got to work in my two new West Wing roles. It was a lot to handle. Moving from one side of the house to the other was like night and day. Everything about the West Wing was different from the East Wing: the schedule, the people, the way it was run. I also saw another side of the president, one I had heard of, maybe even seen on occasion—but never up close and so personal. His mood could change on a dime, and his anger was swift, loud, jarring, vicious, and not always just. Despite a couple of earlier dustups with him when I was Mrs. Trump's employee, I had been spared what it was really like to be on the wrong side of him. I was not prepared for the way he spoke to or yelled at people, which I know sounds crazy when you consider his Twitter account or even his interviews with the press. But when I began to see how his temper wasn't just for shock value or the cameras, I began to regret my decision to go to the West Wing.

Having intimate knowledge of Sean Spicer's and Sarah Sanders's experiences, I was well aware how frustrating it could be to run a communications operation for a president who changed his message by the second, often without telling anyone else what he was going to do. Working as Trump's spokesperson was like sitting in a beautiful office while a sprinkler system pours water down on you every second and ruins everything on your desk—except in this case the water took the form of tweets and words and statements. We were still joking with the press about "Infrastructure Week," as I mentioned earlier, because it was the perfect metaphor for a dysfunctional, broken, chaotic, and frankly crazy system. I could spend days carefully planning an announcement about a new administration initiative on opioids and then wake up at 4:00 a.m. to discover that the president had decided to get into a Twitter fight with Nancy Pelosi, threaten Iran, swear to impose sanctions on China, fire another cabinet member, or respond to some random book author's attacking him (which would always, ironically, make the book a best seller). I can give you endless metaphors to describe the Trump White House from a press person's perspective—living in a house that was always on fire or in an insane asylum where you couldn't tell the difference between the patients and the attendants or on a roller coaster that never stopped—but trust me, it was a hot mess 24/7. How people did the job without going crazy was a question in itself. Maybe none of us did.

Trump, by the way, never understood that he was the one screwing up the messaging. Instead, he would complain to me, "I need a P. T. Barnum!" as his spokesman, just as he would always say, "I need a James Baker!" whenever he was complaining about his current chief of staff. By P. T. Barnum, I think he meant a communications whiz who could somehow charm reporters into writing whatever he wanted them to write. But maybe he just meant he wanted some expert con man. After all, P. T. Barnum's most famous line was "There's a sucker born every minute."

THOUGH IT WAS ALL overwhelming, the first three weeks were the easiest. Thankfully, I had brought my top aide, Annie Kelly, from the East Wing, and I hired a new executive assistant, Baxter Murrell, who basically kept my life in order. With their help I was able to get moving rather quickly, and I could not have done it without them. It was a honeymoon period of sorts, because the press were still impressed by my tussle with the North Korean security guards and there was no pressure to do a briefing immediately. The president went fairly easy on me at first, and I was able to sit in meetings to absorb and learn rather than provide input. Mick Mulvaney and his crew also made things fun and were helpful and understanding with me, especially in the beginning.

But sooner or later, I knew there would be a clamoring for me to appear on camera in daily press briefings at the White House. That was what many people expected the press secretary to do. Six months earlier, the press briefings had been suspended via tweet, as the president had felt that the press was not being fair to him or to Sarah. In their view, the briefings had become vehicles for reporters to get face time on their networks by being as aggressive and confrontational as possible. There was almost never a moment in the Trump White House when the press secretary disseminated information to the public and the press reported it without controversy. Maybe that was our fault, maybe it was their fault, maybe it was both our faults, but in my opinion the briefings weren't doing what they were intended to do.

When I took over for Sarah, that way of thinking was still in place. The president spoke to the press generally every day, and I truly felt that briefings were no longer useful to the public since they heard directly from their president via social media and during press engagements anyway. Kellyanne Conway was one of my biggest advocates and I think understood the president better than I did. We both agreed that he was his own best spokesperson and we certainly didn't need any more stars in our administration. Of course, there was another, more self-serving, reason that I was reluctant to do press briefings: I knew that sooner or later the president would want me to tell the public something that was not true or that would make me sound like a lunatic. That was a lesson I'd learned early on, and I wasn't about to do it unless totally necessary.

I did the best I could to meet one-on-one with every member of my thirty-person team. It was not an easy task combined with all of the other meetings and time spent with the president, but as a manager I wanted to get an idea of what everyone did in their roles, what they thought could be changed, and where they wanted to end up. In my mind as a leader, team morale, empowering people, and the potential for growth were very important to creating a productive and loyal team, so those were my goals. In my own career I've had bosses I learned from and wanted to emulate, and then there were the ones who gave me perhaps the best lessons of all—which was what not to do as a supervisor.

After returning from North Korea, the requests from President Trump to get the press off the complex began. They often varied from one specific reporter to the entire press corps, but it was probably a weekly conversation. For a bit of context, the press were granted access to the White House in the early 1900s. Decades later, President Nixon created a dedicated work space, which became the formal press briefing room. Eventually, small office spaces were given to members of the press corps, and the tents along the driveway of the North Lawn, also known as "pebble beach," were allotted to the networks and cable channels so reporters could do live TV interviews with the White House in the background.

There is no law that says the media must be allowed into the White House, though there is one famous 1977 court case that has been consistently cited—after a reporter whose security clearance was legitimately denied sued for access and won. In short—and they will all disagree—the media are essentially squatters. Why do I know all this? Because one of the things President Trump consistently asked me about was if it would be possible to remove the press corps from the White House. At his behest, I extensively researched what it would take to remove them from the complex, and once it was made clear that we'd likely be sued and lose, I researched different places we could put them other than the press briefing room. Each time the president asked me about my progress on the matter, I let him know I was still working on options, and I always had a new piece of information to share with him so it would look like I was doing something. But for the most part it was the longest (and most successful) stall tactic I have ever used on someone I worked for—who also happened to be the most powerful man in the world.

As I said earlier, Trump's fits of rage were swift, usually brief, but very intense. He had a capacity to find people's weak spot, their vulnerability, and then turn it on them in an incredibly mean, savage, and often effective way. He thought that the worst thing you could be called was a loser or weak, so he deployed those words a lot. He thought that the way to get under the skin of people in the media was to claim that their ratings were bad. He questioned people's confidence, their looks, their intelligence— whatever he thought would do the most damage to someone's psyche. And he was often in a rage about the press, his "unfair" coverage, how no one was ever on TV defending him, and various individual reporters. He really disliked many of them, especially CNN personalities such as Jim Acosta, Kaitlan Collins, and Jake Tapper, all of whom he felt were showboats who made up things to get more TV time.

In his book *Front Row at the Trump Show*, ABC News' Jonathan Karl recounted a story in which President Trump reportedly came into my office at the White House, started screaming at me, and demanded that I throw Kaitlan Collins out of the building. The spirit of Karl's account was right, but some of the facts were a little off as I remember them.

In that instance, Trump had been watching a press briefing in real time and grown infuriated by Collins's line of questioning. He did not come to my office; I'm not sure he even knew where my office was. But he did demand that I walk into the pressroom during the live briefing and "get her out."

"I want her removed from the building now," he ordered.

"Mr. President, I'm working on getting the press out," I replied, referring to my ongoing and hopefully never-ending research into having the press corps moved elsewhere on the White House grounds. "But I can't remove her right this minute."

For someone who was supposedly a press genius, how could Trump not see how disastrous it would look to pull a White House reporter out in the middle of a briefing and then escort her out of the building? If that would help anyone, it would have been Kaitlan, who would be hailed as a First Amendment hero.

Trump didn't like my answer. He wanted her gone and was about to head to the press briefing room himself. He became irate, and it's hard to describe his wrath. It was terrifying for me, for one. "You are weak!" he yelled. "You are a loser! You are useless!" He probably dusted off his "P. T. Barnum" line again, but I was so caught up in trying to endure the rain of brutal insults, I can't say for sure. I did remember his saying I was "the only PR person who wasn't around to do things for him," "no one ever sticks up for me," and "I don't have a fucking PR person who can deal with these nasty reporters."

The president said things like that to pretty much everyone. From my point of view, one frequent victim was White House Counsel Pat Cipollone. If you ever watched the TV show *The Office*, the boss on that show, Michael Scott, had a sworn enemy in the character Toby, the human resources rep who constantly told him what he couldn't do. That was how Trump seemed to feel about the White House lawyers. He didn't like them telling him that things he wanted to do were unethical or illegal. So he'd scream at them. But then he'd usually listen. And then yell at them again later.

Usually we all just took it, and waited for something to distract him or his rage to cool off. Often Dan Scavino, who was usually nearby, would come in and bail us out. "Mr. President, a new Rasmussen poll has you up by two!" Anything that would change the subject and improve his mood.

I'm not sure I'll ever be able to accurately describe all that I did on behalf of the press, nor will I be able to accurately describe the demands the press made on my teams or the frustration we felt almost every day when they seized on pointless stories such as the president stumbling on a ramp or the false and hateful profiles done about members of the senior staff—including me. I will say this—and if even one reporter changes his or her ways, I'll feel successful—there is nothing more maddening than calling a reporter out for inaccuracies and showing proof, only to have him or her say, "I stand by my reporting." The media never seemed to be held accountable when their reporting was wrong, because it seemed if it was at Trump's expense, that somehow meant it didn't matter.

AND THERE WAS, OF course, the particular challenge of working for both him and the first lady—what I'd thought was an ingenious plan to make things run more smoothly. But all too often, balancing the two created an awkward dynamic, especially when they had differing interests. On September 11, 2019, for example, Mrs. Trump was present for a discussion in the Oval Office on e-cigarettes. Alex Azar, the secretary of health and human services, joined the first lady and Kellyanne Conway in urging the president to clear the market of flavored e-cigarettes because the companies were essentially marketing to children. There were others in the room, including the lawyers, staff secretary, and political folks, who cautioned him to slow down, that e-cigs help adults who are trying to quit smoking.

The person who seemed to be really passionate about the topic was Kellyanne. She had shrewdly gone to the first lady long before to brief her and I assume get her to back Kellyanne's position, knowing that Mrs. Trump felt strongly about issues involving children. Kellyanne advised her that the political people were interfering in getting the president on board.

The president, meanwhile, was ambivalent. We had briefed him a couple of times on the issue. If he was talking to Kellyanne, he agreed with her that we should enact a ban on the flavors. If he was talking to his campaign folks such as Brad Parscale he would say we shouldn't do anything because it would piss off the base. Finally, he suggested supporting an awareness campaign to keep children from using flavored e-cigarettes without pulling them from the marketplace. The compromise was apparently backed up by polling given to him by Parscale, which said that "his base" would not like it if the government regulated what adults could and could not purchase.

We kept going back and forth, and, just before the president and first lady were to participate in events commemorating the September 11, 2001, terrorist attacks, we had an impromptu meeting about the subject in the Oval Office.

Mrs. Trump had that look in her eye—she wasn't going to be rolled on this one. She kept pushing for a ban on flavored e-cigarettes and told me to bring the press in so we could announce it right away.

I reminded her that it was 9/11 and she and the president still needed to take part in the moments of silence to honor the attacks. This solemn occasion was not the time to make such an announcement. And frankly, we didn't have our shit together to announce anything but I didn't mention that part. What would be banned? What flavors? How would it happen? When? Look, as a mother myself, I was of the same thinking as Kellyanne and the first lady, but in terms of doing my job, I suggested that we take a couple of days to develop a more strategic approach.

She wasn't having it, knowing that stalling would play into the hands of the ban's opponents. She waved me away. "This is bad for the children," she said. "Call them in," she ordered, referring to the press pool.

I was more than annoyed. She usually listened to me and took my side. Now she was dug in and wanted to get her way. It was an incident that made me look incompetent—again—when we didn't have a comprehensive messaging plan in place. It was another example of one or both of the Trumps' not understanding or caring what it took to do a thing right with the media; they just expected it to be done.

The president must have seen that look in her eye, too. He knew where she was going. "Okay, fine, fine," he said. So he ended up agreeing to call in the press and announced that we would be clearing the market of flavored e-cigarettes in the near future.

A few weeks later, he backtracked after Parscale told him how angry his supporters were. In other words, it was just another day in Trump World.

14

_____

Killers

When God puts you with a bunch of eagles,
don't have a chicken mentality.

—JOEL OSTEEN

In early September, I got pulled into another crazy controversy. A few days earlier, Trump had been briefed at Camp David about a hurricane threatening the southern US border. At one point the hurricane guy told us—and I heard him with my own ears—that it was possible that the storm could change and threaten states like Alabama, but the trajectory was always shifting.

Trump promptly went out and told people that the hurricane was threatening Alabama, along with Georgia, Florida, and other states. Around that time, the projections for the hurricane shifted, just as the hurricane guy had said they might, and Alabama was no longer threatened. The press jumped on that with the usual giddiness: Trump is lying! Trump is misinformed! Trump doesn't know a map! To his detriment, the president refused to admit that the storm projections had changed or that he ever could be wrong. So he doubled down. Alabama was in the path of the storm, he insisted. That had been true—or potentially true—at one point in time. But the president was as stubborn as his wife and never wanted to concede anything to the press.

On September 4, 2019, just before we were to do a briefing to the press on the hurricane, I was with the president in the Oval Office discussing the storm, with a new map sitting on an easel by his desk that did not show the hurricane's path threatening Alabama.

"Why isn't it showing it going to Alabama?" he demanded, still furious about getting all this grief.

Then he took a black Sharpie from the Resolute Desk and drew a line on the map to show us what the original projection for the storm had been. He wanted to show us that he had been right.

The rest of us in the room—I think it was me and Hogan and a couple of others—were over that. We were like, yes, they did tell us that. You were right, Mr. President. The press isn't being fair. Blah blah blah. We just wanted to move on.

At that point, the president's body man came in. "Sir, the press is waiting for you." Trump was already behind schedule, so we all scurried to move things along.

"All right, all right," he said. "Let's bring them in," and when he makes that demand, people move quickly including the wrangler who is tasked with bringing them in efficiently.

No one thought—I certainly didn't—about the fact that the map was still displayed for the press to see.

When the reporters came in, some of them zeroed in on the map. The story was too good to resist: The president was creating his own reality again! He was doctoring official documents! Even though, in that case, Trump actually hadn't planned to show the map to the press. On the other hand, he refused to explain anything, admit he was mistaken or ever could have been wrong. Sharpiegate, Trump scandal number 4,326, was born.

No reporter believed our explanations by that point. They believed that everyone around the president was a liar. It was yet another example of a fascinating interaction between Trump and the media; in my opinion both were wrong in the way they behaved, but neither cared. That created a toxic and in some ways twisted relationship between Trump and many of the reporters who covered him that in a weird way both hurt and benefited each side. The truth was that Sharpiegate was more our fault than Trump's. But as always, the president proved to be his own worst enemy, refusing to admit any possible error on anything that might have killed the story before it ballooned and then using the latest media attacks on him to rally the support of his base.

WE WERE JUST GETTING past Sharpiegate when, on September 10, 2019, John Bolton and the president had a parting of the ways. As it happened, I was walking into the Oval Office just as Bolton was walking out. He said, "I'll see you later" to me, which was a surprise, as he hadn't said a word to me since the Mira Ricardel incident. But the way he said it made it clear to me that he was gone for good.

Bolton left a resignation letter on every senior staffer's desk, including mine. I assumed it was because he knew it would leak and he wanted to leave on his own terms. On White House letterhead in envelopes from the NSC marked "Personal and Confidential," it read, "Dear Mr. President, I hereby resign, effective immediately, as Asst to the President for National Security Affairs. Thank you for having afforded me this opportunity to serve our country."

As soon as the president found out about the letter, we scrambled with Dan Scavino to get a tweet out saying that Trump had asked for Bolton's resignation. According to the president, he had dismissed him the night before. "Bolton was a nut job," he said. "I got him out." For the rest of my time in the White House, the president would describe his former national security advisor as unstable, crazy, dangerous, or wanting to start a war. (He also used to say that John Kelly "really sucked.") That was what Trump did to anyone who left on unsatisfactory terms. It never occurred to him that perhaps that reflected on his hiring practices or that months before he had been singing their praises. In any event, national security advisor number three was out the door.

Not long after that, I accompanied the president to the United Nations, where he would have a series of one-on-one, or bilateral, meetings with various foreign heads of state. Bolton's departure was among the frequent topics of discussion, along with a brewing scandal over a country most people hadn't thought about in ages, if ever: Ukraine.

BEFORE WE GET TO that scandal, which of course led to Trump's (first) impeachment, let me take a step back and try to describe what it is like to sit with Donald Trump and foreign leaders. One-on-one, or bilateral, meetings are a key part of the presidency. They most often take place at large gatherings such as the G7 or G20 and have a basic formula. The two world leaders meet and sit in two chairs at the end of a room. Flanking each of them are generally ten members of the delegation, often including the secretary of state, the head of the National Security Council and staff, cabinet secretaries, and a press representative. The meetings generally last around thirty minutes and consist of pleasantries and introductions in the beginning; then each country's press pool is allowed to come in for a few minutes for a photo opportunity and perhaps a question or two. Of course, I was there when our president kept the press in the room for easily thirty minutes or more, while the other world leader just sat there, mostly silent. Once the press are dismissed from the room, the two leaders talk about the issues facing each country and concerns that either of them may have. Depending on the country, the meetings can be kind of boring for some, very productive, heated, or hilarious; usually with Trump it was more than one of those at a time.

Having never attended a bilateral meeting before that job, I have to say that for the most part I thought President Trump was effective at them. He was charming and funny, but he also stood very strong and truly kept to his mantra of "America first." In almost every meeting with our NATO allies, he brought up the fact that the United States paid far more than any other country in paying to help defend other nations, and he was not shy about letting leaders such as German chancellor Angela Merkel, Chinese president Xi Jinping, Canadian prime minister Justin Trudeau, and Japanese prime minister Shinzu Abe know his opinions. "We're being ripped off," I remember him telling Italian prime minister Giuseppe Conte, "but that's not an elegant way to say it." Trump didn't seem to be a big fan of European leaders in general. Of French prime minister Emmanuel Macron, a trim, soft-spoken man, Trump scoffed, "He's a wuss guy. He's all of a hundred twenty pounds of fury."

One of the funniest meetings was with British prime minister Boris Johnson, one of the few European leaders Trump seemed to tolerate. Conversations between those two, both pudgy white guys with crazy hair, redefined the word *random*. Johnson once told us over breakfast that Australia was "the most deadly country—spiders, snakes, crocodiles and kangaroos." Then they discussed how powerful kangaroos were at considerable length.

Trump commented on entrepreneurs such as Elon Musk who were working on rockets. "Rich guys love space," he observed.

Then they talked about some political figure who'd just had surgery, which they thought had involved the removal of a gallbladder. "Can you put a new gallbladder in?" Johnson asked, chomping away on scrambled eggs and sausage. "I don't know what a gallbladder does."

"It has something to do with alcohol," Trump replied.

When we traveled to Davos, Trump marveled about the country of Switzerland, which was clean, orderly, and full of rich people—very much his kind of scene. President Germophobe gushed to Swiss president Ueli Maurer, "Everything here is so clean. My hotel room is spotless." Then he complained about wind power and warned that a bunch of windmills could ruin the Swiss scenery. "Can you imagine how ugly it would be with your mountains? So beautiful. Also kills all the birds." When the Swiss president spoke proudly about the success of the country's vocational programs for unemployed adults, Trump asked, "Do you teach them how to make watches?" He turned to the rest of us, helpfully adding "They have great watches."

He complained about the head of the European Union, Jean-Claude Juncker, "China is easier to deal with than the EU. . . . Jean-Claude, he's a real beauty." Then he said of the Iranians, "They're always trouble. They're wired that way, perhaps."

There was a running joke about me in the White House that although most staffers were excited to meet figures such as Her Highness the queen of England, I liked the meetings with dictators best. I want to be clear that I did not and do not admire those people. I just found the policy and national security conversation that surrounded them fascinating, and the way our president talked to them—well, at times, it could be unique.

Trump seemed fixated on dictators, too. It almost seemed as if he admired their toughness and aggression, but he also was genuinely freaked out about nuclear war. Naturally, as POTUS, he received frequent briefings about nuclear strikes and their potential impact and was (rightly) scared straight about the dangers of getting into a war with one of those guys. He said many times that nuclear war was his biggest worry. "Forget climate change," he once told me. "What we have to worry about is the bomb."

He always seemed to want dictators to respect him. It looked to me like he tried everything to get them to take him seriously or to at least get along. That seemed to work with some of them. He and Xi, despite the complicated issues and rhetoric between China and the United States, seemed to get along pretty well. Even Kim Jong-un cracked a smile once or twice. But there was one who was a tough nut for Trump to crack: Vladimir Putin.

Putin was handsome in a "power is an aphrodisiac" sort of way. Very confident. Very cool. Very unflappable. And it wasn't as if there were a bunch of male models out there running the world to compare him to. I don't know if it was because of all the media attention the US press gave him or the fact that he seemed to be proud to be an allegedly coldhearted killer, but I was fascinated by him.

My first encounter as press secretary with Putin was in Osaka, Japan, at a bilateral meeting during the G20 in 2019. That day I was seated next to Fiona Hill, our top Russian aide at the National Security Council. Everything started out as it always did, all of us taking a moment to provide introductions and the two leaders chatting for a couple of moments before the president would look at the other leader and ask if he was ready for the circus and the chaos of the press. Trump would usually spend a minute or two talking about what "animals" the reporters could be, and then the press would come in. With President Putin, Trump started out the same, then changed his tone.

With all the talk of sanctions against Russia for interfering in the 2016 election and for various human rights abuses, Trump told Putin, "Okay, I'm going to act a little tougher with you for a few minutes. But it's for the cameras, and after they leave we'll talk. You understand."

I was surprised by that, but no one else on our side seemed to be, so I thought maybe it was normal. It was my first bilat with him, and for all I knew the NSC had prepped him to say certain things. I thought it made him look weak and undercut whatever criticisms he had of Russia.

Putin responded to Trump's comment calmly. He never seemed to be charmed by Trump or even impressed by him. If anything, the Russian seemed to look down on him. I can only imagine that must have irritated Trump, making him want to earn Putin's respect even more. Putin probably knew all that. He was an old KGB guy and renowned as a master of head games. And Trump was a very easy mark. It struck me that Trump seemed to really want to impress Putin. And I think Putin knew that.

As the meeting began, Fiona Hill leaned over and asked me if I had noticed Putin's translator, who was a very attractive brunette woman with long hair, a pretty face, and a wonderful figure. She proceeded to tell me that she suspected the woman had been selected by Putin specifically to distract our president. I was fascinated and from then on couldn't stop watching for any interactions between the interpreter and President Trump. Sure enough, he addressed her directly a couple of times, jokingly and casually, but it was certainly nothing I hadn't seen him do before.

As the meeting progressed, I noticed Putin coughing on several occasions. Well, actually it was a little more subtle, more like small cough sounds accompanied by throat clearings. He did it so much that it became distracting to me. I thought to myself, "Dude, drink some of that damn water sitting in front of you!" Finally, I said something to Fiona about it, and she quickly replied, "He's probably doing that on purpose—he knows full well the president doesn't like germs."

I was blown away; why hadn't I thought of that? Actually, if I had done what Putin was doing, Trump would have freaked out and yelled at me to get the hell out of the room, probably fired me. In that instance, he didn't say anything. But I have to imagine that Putin was messing up his head big time.

ANOTHER MEMORABLE DICTATOR WAS Turkey's Recep Tayyip Erdoğan, who we met with at the G20. While the main topic of the meeting was Turkey potentially purchasing some aircraft from the United States, the conversation quickly turned to a missile deal done under the Obama administration. Trump looked at our delegation, pointed to Erdoğan, and said, "This is no innocent baby, but he got screwed here."

Every time he talked to the Turks or about Turkey, Trump had the same talking point: how they'd been screwed by Obama. Trump loved to order cabinet secretaries, staff, or whoever else was sitting near him to give this dictator or that whatever it was he wanted. I think he enjoyed seeing his own people squirm when he said things such as "Let's give it to them. Let's get it done. Now." He would generally follow up with said dictator with a casual "If it doesn't happen, you just call me, you don't need to call the others, I will take care of it." I believe he must have thought it made him look tough and powerful. As for us? We generally slow walked or ignored the president and very rarely did exactly what he asked.

Another entertaining aspect of those meetings was that because the president got bored quickly, he would move on to different subjects, often using as a segue "Let's talk about another subject, that is too unpleasant for me." Or he'd say, "We can't talk about trade now. Boring."

At one point in the bilat, a member of Turkey's delegation asked for a private meeting with our president and Ambassador Bolton, who had not yet left the administration. The president, as he did often when referring to John Bolton, pointed at him and said,

"There he is. Congrats. Except he wants to attack every country in the world. You don't want Bolton, he'll use you as a target." I wasn't sure what the "congrats" was referring to, but I could tell that Bolton didn't love the rest of the statement.

It was during the same meeting that the president stopped and looked at me. "Stephanie, who's tougher?" he asked me. "Xi or this guy?" He looked over at Erdoğan, who smiled.

Everyone turned my way while I tried to think of a good response. I mean, WTF is a person supposed to say to that?

Erdoğan was staring me down, and I truly didn't know which answer would be more offensive—that he was or wasn't "tougher" than the Chinese president. So I stuck with "I don't know, sir, he looks pretty tough," pointing to the Turkish president.

But the highlight of that particular bilat was when the subject of the film *Midnight Express* came up. The 1978 movie is based on a true story about a young American who was caught smuggling drugs in Istanbul and thrown into a violent Turkish prison. The film was criticized at the time for being over-the-top anti-Turkish. The *New York* magazine critic said that all the Turks in the movie were "presented as degenerate, stupid slobs" and even "sub-human." I don't know if Trump was trying to put Erdoğan off his game or if the movie just popped into his head. But at one point he looked at the Turkish delegation and asked, out of the blue, "Have any of you seen *Midnight Express*? That's a dark movie for you guys." There was little reaction from the delegation, maybe a few polite chuckles, before the conversation moved on, as if the president of the United States hadn't just blurted that out.

After the meeting, Trump came up to me and corrected my answer to his on-the-spot question about Erdoğan. "Xi is actually a lot tougher, I've seen him in action, he's a stone-cold killer." Whatever that meant. Maybe I didn't really want to know.

With Prime Minister Imran Khan of Pakistan, Trump talked

about the Taliban and terrorism, the tension between Pakistan and India based on religious differences, and Pakistan's issues with Afghanistan. Prime Minister Khan told Trump, "There is a beginning of a crisis between my country and India. We look to the United States to put out flames in the world."

Trump responded to that impassioned plea. "I trust this gentleman right here, I trust Pakistan," he said. "I would say, as somewhat a student of history, there's almost always a solution."

Trump then spoke about our potential trip to India and the discussions he had had with Modi about toilets and his amazement that so many people would use the bathroom in the streets. He couldn't get over that. India reminded him of California with all of the homelessness, he said.

Khan kept pressing his case about India. At that point in the meeting Trump waved his hands and looked around the room, seeming somewhat annoyed and maybe even bored. "Doesn't seem appropriate to talk about trade now. Boring, very boring."

To Egyptian president Abdel Fattah el-Sisi, Trump complained about the Egyptians buying weapons from the Russians instead of the United States. The Egyptian president blamed the Obama administration for its decision. "This is a consequence of Obama's administration," he said. "I don't wish to speak on this issue."

Trump was insistent that he could get Egypt a great deal from the United States. "I'll approve things myself, and you'll get everything very quickly," he said. "I don't want you paying billions of dollars to Russia." Like I said earlier, President Trump really fought for our country in the meetings. "America first" was not just some slogan.

The Egyptians then said that they were ordering planes from two other countries with which they had "a great relationship." That led Trump to remark, "Japan said they were our friends right before they bombed us."

Then the talk turned to the Nile River and a problem involving Ethiopians building a dam funded by the Chinese. Trump didn't seem to understand the issue. Instead he turned the conversation to the Olympics. "This is why they win the gold," he said, referring to the Ethiopians (I think). "They run up and down that river."

Trump told Iraqi prime minister Mustafa al-Khadmi that he had watched a documentary on the wives of ISIS. "They are stone-cold mean," he said.

And so on.

ON SEPTEMBER 22, 2019, I traveled with the president to the United Nations, where he was to address the General Assembly with a major speech on religious liberty, calling attention to the 80 percent of the world's population that was persecuted in some way for exercising their religious beliefs. You would have thought that the United States' retaking global leadership on an issue that affected billions of people would have commanded at least some attention. But all anybody in the press really wanted to talk about was what Trump was calling "a perfect phone call."

_____

15

Impeachment Number One

Some people can't function without negativity because
bringing down others makes them feel better.

—UNKNOWN

On July 25, 2019, President Trump had a telephone call with the newly elected president of Ukraine, a former actor named Volodymyr Zelensky. One of the topics the president raised was the actions of former vice president Joe Biden and his son, Hunter, who had been on the board of Burisma, a Ukrainian gas company. In August, a whistleblower, thought to be someone in the intelligence community who had not been on the call but had heard about it from others, made a formal complaint, alleging that "the President of the United States is using the power of his office to solicit interference from a foreign country in the 2020 U.S. election." On September 20, the media brought that complaint to light.

The Democrats pounced immediately, and the story hung in the air over our preparations for the New York trip. The left wing of the party pushed House speaker Nancy Pelosi to call for Trump's impeachment right out of the gate, but she seemed hesitant to go too far too fast though she was clearly moving in that direction, which put the president even more on edge.

In early August, two horrific shootings had happened back to back in our country. The first was at a Walmart in El Paso, Texas; the second was at a local bar in Dayton, Ohio.

We visited both states on August 7. The plan was to spend time with the victims and families and also make stops to thank the first responders. The trip was a disaster for me professionally. The president and first lady were both angry with me when there was no press in the intensive care unit in Ohio to capture all of the medical staff clapping and cheering for them and taking selfies. I tried to explain that we weren't allowed to have the press in areas where there were patients, for both privacy and concerns over keeping the environment as sterile as possible. Didn't matter. He was angry, and to my dismay, she was in total agreement with him. As I've mentioned I always expected her to have my back, which was not realistic, but she wasn't cutting me any slack. I started to wonder if maybe she was starting to resent the fact that my loyalties were divided; maybe working for her and her husband wasn't such a good idea after all.

At every stop after that, the president kept raising the issue with me. "This first responders control center is nice, but I can't believe you didn't have them—meaning the press—in the hospital. What a waste." After trying to explain the logistics and reasoning to him once again, I realized that it was no use. I would just have to take the criticism for as long as he dished it out.

As we were taking off from Ohio to Texas, my stomach dropped when I saw that the two local Democratic politicians who, at our invitation, had spent the day with us on the steps of the hospital we had just visited were holding a press conference. They had clearly preplanned it, and I watched the president's mood torpedo as he watched it unfold on television. They said that Trump's visit had been a publicity stunt and had done nothing to heal the community. The president then turned to me and, for the first time since I had met the man, totally unleashed on me. His eyes were fixed and filled with anger. His face had passed red and was turning almost purple.

"Where the hell are our people? Why are those two on TV right now and there is no one to defend me? Why are you even on this plane? What do I have a whole team of people for if there is no one fucking defending me? And I am stuck on this fucking airplane and can't do anything!"

It's impossible to explain how that felt, sitting there defenseless, while the president of the United States unleashed his "fire and fury" on me. And in front of others, too, who I'm sure were just relieved that it wasn't them, because trust me, I'd been on that end of things too. They included my girl, the first lady, who sat watching it all go down and, worse, seemed to fuel it.

I said, "We did have a really good photo op with law enforcement officers," spinning desperately and, frankly, badly.

The first lady wasn't having it. "No. No. There should have been press at that first stop. That would have been better. It didn't look good." That rattled me, because she'd never talked to me like that before. She was almost egging him on after hearing him yell at me.

As they both continued their verbal assault, I was pissed, particularly at Mrs. Trump. I felt let down by her—not that she had to always defend me or could never criticize me, but in this case she had to know that my explanation was logical. I knew she knew that what I was saying made sense. And that the president's fury felt over the top and abusive. Was she joining in on the attack on me because she was truly pissed off, or did she see where it was going and didn't want the president taking his anger out on her, too? I don't know. I never found out. She and I never discussed it again.

Some who read this might ask, "Why did you put up with

that?" I was working for the president of the United States. We all put up with it. It never occurred to me then that I had any other choice, and to be honest, I felt like I was being a baby most of the time. And the storms always passed eventually. It was a cycle that I think we all as staff encouraged with our silence.

The president referenced the group of reporters who were traveling with us on Air Force One. "Go back there, go back right now and tell them how much everyone in that hospital loved me," he said. "Go, do it. Do it now." I was shaking and nervous. He had yelled at me before—he yelled at everyone—but I'd never experienced that kind of pure rage from him before.

I left the office on Air Force One and went into the staff conference room to gather my bearings and pull myself together. I had someone call Ohio senator Rob Portman and see if we could get him on TV to talk about what a great visit it had been. Then I tried to think of what I would say to the pool in the back other than "Medical staff in the hospital loved the president," which the press would have just laughed at. What would P. T. Barnum do? I didn't go back there. If I had done what he wanted—gone back to the reporters on a day when we all should have been focused on shooting victims and their families to say, "Hey, guys, just want you to know the hospital staff really loved President Trump"— they would have made a fool out of him, and of me.

I stayed out of sight for the rest of the flight, hoping that the Texas visit would go better than the Ohio visit and all would be forgotten. I had my team call ahead to see if we could get some friendly local Republicans to speak to the press as soon as our last visit concluded. Thankfully, the visits in Texas went well despite the tragic reason we were there, and though I gave the press access to everything that the president and first lady did on that leg of the trip, he continued to tell me that I had missed a real opportunity in Ohio. At one point the president, confident that he could do what I could not, spoke with the White House press pool and let them know that the medical staff in Ohio, "covered in blood from surgeries," had crowded around him for pictures. That, predictably, led reporters to search social media for photos from the hospital. No one was covered in blood, as Trump had said, which became another terrible story about the president's lies.

I WOULD RECEIVE A second ass chewing like that from the commander in chief on Air Force One, but on a different trip and over the impeachment hearings. I remember that we were headed home from an event and after the hearings had concluded for the day, the president once again questioned why "no one" was on TV defending him. I was sitting in the chair right next to his desk, and Dan Scavino and Jared Kushner were on the couches across from me. The president's voice kept getting louder and louder as he yelled, "No one ever defends me! Where are your managers, Stephanie?"

I gave Jared a pointed look because he had forced me to hire someone to "handle" impeachment messaging a couple of weeks earlier; more on that in a bit. I followed up with "Sir, I will text them right now, they are on the Hill and should be getting people out to the cameras" and started to frantically text one of my deputies. The president didn't give one shit as I kept reading him the texts I was receiving explaining that the media were taking only Democrats and the White House staff were doing all they could to push the likes of representatives Jim Jordan and Mark Meadows to the cameras. In the president's mind, the day had been a total waste and I was a complete failure for not getting more people onto TV to defend him. He reiterated that he had an entire communications team that was pointless and that before me "he always got the best press." Right. At one point he got on the phone with Jim Jordan, who seemed to calm him down for about ten minutes; then the coverage riled him up again. Meanwhile, Scavino and Kushner sat there, not really saying much to help, though Scavino kept offering up positive things he was seeing on social media. This was yet another time that Jared's habit of taking over or telling staff what to do, in this case me, was of no help at all and only caused trouble for the idiot staffer (me) who had done what he directed. Adding salt to the wound of course was watching him sit there and watch me get my ass kicked up and down the plane while not offering up any of his part in my "shitty managers" who were on the ground.

BACK TO THE BEGINNING of impeachment number one. On September 24, the president spoke at the UN General Assembly. He walked into the heart of globalism and declared, "The future does not belong to globalists; it belongs to patriots." He called out China for unfair trade practices. But none of that really mattered because that same day, House speaker Nancy Pelosi announced a formal impeachment inquiry related to the Ukraine allegations.

The White House had known about the whistleblower's complaint for only a couple of weeks, and to my knowledge the White House Counsel's Office hadn't seen the transcript of the notorious phone call until only a few days before the story broke. They had not shared the details of the call with anyone other than the president. Once the inquiry was announced, an internal debate about releasing the full transcript to the public began. As a comms person, I could not weigh in because I had not seen the transcript. Since no one trusted anyone enough to share information, I was already one hundred steps behind. I knew enough about Trump to know that he could have said anything on that call. But he was usually pretty careful not to threaten anyone he was talking to one-on-one, as the whistleblower was alleging.

We were in a hold room at the United Nations, waiting for the president to finish a bilateral meeting, when I saw the transcript of the Ukraine call for the first time, thanks to Vice President Mike Pence, who handed it to me. I guess he thought it was absurd that the communications director, who would have to answer the questions about this from the press, hadn't even seen it. I recall that Mick Mulvaney was also in the room.

I read through the transcript quickly. To be honest, nothing jumped out at me as cause for alarm. And I want to explain why, because I think it helps explain the mindset of some of us who worked in the Trump White House. In the first place, I certainly didn't see any smoking gun in the transcript that Trump had committed an illegal act. But also, as the last chapter revealed, the president frequently said insane things to foreign leaders. Sometimes they were just silly or offensive, sometimes they were offhand remarks that would inadvertently upend the carefully crafted policies of our diplomatic and national security professionals, sometimes they were sheer bluster. The point is, over time we all had grown numb to the broad range of things he said. So when I read the transcript, placing his remarks to Zelensky on the sliding scale of Trump language, it didn't rate very high in my opinion. His behavior had become normalized. As his press secretary, I accept responsibility for my role in this normalization because it was my job to deny, deflect, or play down his comments to the press. Maybe, too, I was just too deep into the spin at that point myself. We were operating in a state of siege mentality, conditioned to see the press as the enemy. Looking back, it's hard to point to actions I could have done to change his behavior. Resigning in protest? Ask generals Mattis or Kelly, even Ambassador Bolton, if that had

had any effect on how Donald Trump operated. And again, at that point how would I ever find another job? I was not like a vast majority of the "originals," who didn't have to worry about their finances. I can't speak for my colleagues, but I know that I began to feel trapped. Because however much I agreed with the president's policies and hated the unfair attacks by the Democrats and some in the media, I started to wonder what part I was playing in normalizing his behavior and the future of our party. What was worse, was I also sacrificing my own integrity and betraying my moral compass? Probably yes.

So for all those complicated reasons, the call didn't bother me at the time, and Mick and I were of the opinion that we didn't need to release the transcript right away—not because we had anything to hide but because we wanted time to prepare a proper communications strategy (I hope you the reader are seeing a pattern here). Mick placed a call to Pat asking about the transcript and Pat replied, "You guys don't need to worry about that. Bill (Barr) and I have a communications plan in place."

As you can imagine, that floored us—and infuriated me. Nothing ever worked the way it was supposed to in the Trump White House, so on one level it shouldn't have been surprising. Out of mostly just frustration I asked Mick how it was possible that as both press secretary and director of communications, I hadn't been looped into the conversations on an issue that was already dominating every news channel in the United States and soon the world. Mick agreed, reminding me that he was chief of staff and hadn't been looped in, either. His outrage was even greater than mine. How can you serve as chief of a staff that doesn't clue you in on a major scandal? We asked that question of Pat, who casually let us know just what we could do with our outrage. "This is a legal issue," he said dismissively. He and the Attorney General would handle it.

Having worked for the Arizona attorney general years before, I fully understood the dynamic that was unfolding. Lawyers just want to win the case, and they are trained to be cautious about sharing information. Meanwhile, communications people want to win in the court of public opinion, which means giving people accurate and honest information but in a strategic way. Often, those two priorities clash, but there is usually time to negotiate a middle ground.

In that case, there was no time, and tempers were already starting to flare, especially between Pat and the increasingly irate Mick.

"How the hell do we not know about any of this?" Mick rightly asked. "We've got to know what's going on so we can be prepared for a response."

Pat would say things such as "Okay, sure" and then ignore him. Or, when pressed, he would say that we couldn't possibly understand the legalese of the case. One of the problems with the revolving-door cast of characters that constituted the Trump White House was that nobody had time to build trust in others. Who could blame Pat, in a way? Mick was chief of staff number three. I was press secretary number three and communications director number five. We all knew that there could be numbers four and six at any moment. So why should Pat tell either of us anything? We all looked out for ourselves.

After Cipollone basically told us to stuff our complaints into the nearest trash can, the president and first lady walked in. The boss was taken to another quick pull-aside meeting in that room with German chancellor Angela Merkel, so I filled Mrs. Trump in on what was going down. She agreed that we should wait to get all our ducks in a row before rushing to release the transcript—no surprise, since that was generally her attitude with the press—but she clearly wasn't the decision maker in that instance.

The president joined us a few minutes later, and the three of us sat together in nearby chairs to discuss the issue. I told the president that I saw no upside in releasing the transcript right away because we weren't yet prepared and people would read into it whatever they wanted to. He disagreed and wanted it sent out immediately because he had done nothing wrong. It was around that time that he convinced himself he had made a "perfect phone call." I was so used to watching him spin whatever he wanted to be true into reality that it didn't catch me off guard at all. He was always innocent of every accusation, always unfairly targeted, unfairly treated. That was par for the course with him.

"Yes, sir, I understand and agree with you," I said. "But the Dems have already seen it and have announced an inquiry, so their narrative is set that you did something wrong." I told him that we needed to find out what parts they had issues with so we could understand it, then put a communications strategy together. Put a strategy together? Think things through? Work the press? That sounded like a waste of time to him.

He wasn't persuaded, even when Mrs. Trump chimed in. "Donald, there is no rush," she said. "Let us get through UNGA, and then you can decide."

That at least gave us a reprieve from making the decision then and there, and I hoped we'd have some time to further review the full transcript and get some context around a few of the exchanges. Trump didn't say much to that, but he clearly wasn't listening.

I found out later that night via Mulvaney that the decision had been made to release a partial transcript the next day, prior to the president's bilateral meetings and end-of-day press conference. The thought behind that decision was that it would preempt the questions he would get from the press throughout the day. I disagreed. Releasing a partial transcript would be the worst possible strategy; it would look as though we had something to hide.

I called Pat and let him know my misgivings, telling him that a partial transcript would only make the press ask "Why not all of it?" and seek out what was in the part we didn't release. He suggested that I connect him with some of our more friendly reporters. Friendly reporters? I wished I knew who they were. He said that we could "give them background information off the record." I didn't bother to tell Pat that "background information off the record" would give us absolutely no results or goodwill with the media because if it is off the record it can't be reported. As I said a few chapters back, everyone thinks they're a comms expert.

SOON AFTER, I WAS with the president when he met with the Ukrainian president in person. Before the meeting began, Trump noticed that he hadn't gotten his customary soda. "Why no Diet Coke for me?" he asked. Then he turned to President Zelensky. "It's probably because everyone would call it a bribe," he said with annoyance.

They briefly touched on the now-infamous phone call. Trump asked Zelensky, "Can you believe all the hype about this?"

Maybe he was trying to court Trump's favor, but Zelensky seemed visibly irritated, telling us that there had been nothing to the call. Moments later, he told the press the same thing, that the call had been only about arranging a meeting and changing Washington's tone on Ukraine, and "not linked to Burisma or military aid."

When we returned to DC, the arguments over communications strategy continued. It bears repeating that in a traditional White House, the comms team would implement a strategy with input from the counsel's office, the chief of staff, and obviously the president. In the Trump White House, it was another free-for-all, with the president generally going with the advice offered by the last person he had talked to.

I think those who work in communications will sympathize with the fact that people always think they are experts on communications. They think it's easy to "spin" the media and come up with the perfect statement and that they generally can do the job better than you can. If a person or organization gets bad press, it's the fault of the communications shop—that was certainly always the thinking of the president and his chief of staff, Jared. If they get good press though, you get no credit because it was so easy. Many people in the Trump administration suffered from that affliction, from the president on down. So it was nothing new that Trump and Rudy and AG Barr and Pat and Jared and everyone else with an opinion anywhere in the White House thought they knew how to handle matters best.

ONE DAY, IN THE midst of the growing crisis, the president called me into the Oval Office and suggested something that made my stomach drop. He was sitting at the Resolute Desk, transcript in hand, when he said, "Stephanie, I want you to go out to your stage tomorrow and reenact the phone call. It was absolutely perfect." The stage he was referring to was the podium in the James S. Brady Press Briefing Room. "Okay, no problem," I replied. "I'm not sure what exactly you need me to do, though. Read the transcript to the press?"

He was irritated by my confusion. "No. You need to act it out. With voices. You need to really reenact the phone call so people will understand how perfect it was. Elegant, really. So you'll need to use two voices and act it out."

Oh. My. Dear. Lord. I was speechless. So for my very first press conference, I was to make my debut in the historic James S. Brady Press Briefing Room and read the transcript to the reporters and cameras, presumably with a Ukrainian accent for Zelensky's portions? And what voice would I be using for Trump? The president grew increasingly animated about the idea, genuinely seeming to sell himself on it. It would fix everything. A million things blew through my head as I said, "Okay" and walked out the door. I spoke to no one as I headed to my office, where I sat down and put my head in my hands. This was it—this was going to be my ticket to being satirized on *Saturday Night Live*—and not in a good way. I would be all over the news, sounding like a straight idiot that night and for years to come. I can barely chronicle the myriad of feelings that went through me for the next hour as I struggled to find a way out of it. I even considered calling in sick. I finally told Mulvaney's deputy, Emma Doyle, about the president's amazing idea (I hope you sense my extreme sarcasm), and, thank the stars for her, she came up with a brilliant plan. She suggested we tell the president that although his idea was a really good one, we should instead ask a member of Congress to read the entire transcript on the floor of the House so it would be copied into the record for history. Different voices optional. We waited until he was about to leave the Oval Office to board Marine One before we sprang the idea on him, calculating that he'd be in a bit of a rush and not so focused. Emma laid the new "strategy" out, and sure as shit, he loved the idea of it being read into the record for all of posterity! A man with an ego will always choose being immortalized in the history books over a press secretary making a fool of herself on her "stage." The next day, one of our most reliable "yes" guys in the House, California congressman Devin Nunes, read the full transcript before the House Intelligence Committee so the perfect phone call could be remembered forever.

OVER THE COURSE OF the next two weeks, tensions between my team and the counsel's office increased. Since Ukraine and impeachment were now a big issue for the president, it wasn't long before Jared inserted himself. There was a time when everyone in the White House would have loved to be on Jared's radar because it meant they mattered. But by now I had figured out that Jared Kushner at your door was sort of like a visit by the Grim Reaper—he always brought trouble and escaped without a scratch. He came into my office one day to ask "where we were" on the messaging for impeachment, citing some concerns that my teams weren't being "aggressive enough" in pushing back. He said it the way he said everything: low key, quiet, as if he were on your side, trying to help.

A back-and-forth ensued, with my trying to explain to him the problems that my teams were dealing with, notably trying to wrestle the most basic of information from the counsel's office. Jared let me know that he would speak to the counsel's office and explain to the staffers there the importance of sharing information. Good luck with that, Jared. They would tell him, "Sure thing" and then ignore him.

About a week after our first conversation, the Slim Reaper came back into my office with a new idea. "Hey, Steph," Jared began in his quiet "I am on your side" voice. "What are your thoughts on putting a team together that is dedicated solely to the impeachment inquiry? It would include members of your press and comms teams, counsel's office, and of course myself and members of my team from time to time." He wanted to have a war room so the team could meet and strategize.

I reminded him that the president had made it clear that that was a nonstarter. He had told all of us that there should be no appearance of a team being put together, and no war room would be needed since he had done nothing wrong. In his mind, if we made a big, visible effort to defend him, he would look guilty. The real problem, of course, was that even if we assembled the finest communicators in the land and put together talking points that would make Rachel Maddow proclaim Trump's innocence, it wouldn't make a bit of difference; Trump would go out and tweet whatever he wanted, or he'd let Rudy Giuliani go say something bonkers.

Anyway, Jared was here to save us and didn't care about any of that. He offered his usual wink and nod. "Let me take care of the president." He loved to brag about how easily he could control his father-in-law, as though he were the ultimate Trump whisperer. I was like, whatever, dude. I don't have time to deal with these daddy issues.

He then gave me a directive that I believe was the start of the true deterioration of our relationship. He told me that I was to hire a certain person to help with comms. I immediately bristled at the idea and said that adding people would only further hurt morale. I explained that if we had access to the information we needed, we would be able to create and implement an effective communications plan. I also explained that in a White House filled with leaks, adding someone new would only exacerbate things. The person Jared had chosen was known as a leaker to some of his colleagues and those on my staff, and I rightfully predicted that if my deputies caught wind of his coming on board, they would not be happy.

"Steph, we're not here to worry about hurt feelings," Jared replied. "We are here to work."

To which I replied, "I agree, but we can't work without information; I'm not sure how adding people corrects the actual problem." I found his comment offensive and more than a little condescending. I was talking about the morale of the team and the dangers of adding more people on top of others already trying to do the same job.

Jared was feeling his oats, as they say on the farm. He no longer struck me as the friendly, politically naive guy of the early Trump years. Now he seemed all-powerful and entitled—a dangerous combination for anyone at the center of power. Emma Doyle and I put our heads together and suggested Pam Bondi, the former Florida attorney general, instead of Jared's choice. She was well known, was great on TV, and had legal experience, the president trusted her, and, most important, the team would trust her. It's not as if we had a ton of options for people wanting to come in and defend Donald Trump anyway. We ended up with both of them.

After my conversation with Jared, I went to the president to express my concerns about being ordered to hire additional people.

Trump replied with the most honest remark I may ever have heard him utter: "You probably don't want to get on Jared's bad side." I'm not sure the president necessarily thought that Jared was a genius. Or that he thought that Jared was going to fix all of his problems with a wave of his hand—no one could. In my view Trump gave Jared basically unlimited power for one reason only: Ivanka. He probably didn't want to tick off his baby girl. That was a reminder to me that no matter how close I might feel to the Trumps, I wasn't family. And family trumped all.

I recall only one occasion where I witnessed the president really push back on Javanka. I was in the Oval, waiting to go over interview requests for the president, while Ivanka was speaking to her father. His assistant came in to remind him that he had only thirty minutes before his next event, which involved criminal justice reform, one of the few legislative programs that had broad bipartisan support. That information did not please the boss. "Tell Jared no more," Trump told Ivanka. "All I ever do are these criminal justice events, and they get me nothing. I'm serious, tell him I am really not happy about this and I'm not sure his advice has been so great." Ivanka did—gently—try to push back in defense of her husband, but the president wasn't having it. "Just tell him, I'm tired of it." It was probably the angriest I'd ever seen him toward his beloved daughter. She wisely didn't push him further and left the Oval, leaving me twenty minutes with a very grumpy boss. Thanks a lot, Ivanka.

In general, the president would often express his reluctance to do "Jared events," leading Jared to complain at one scheduling meeting that we needed to stop saying that things were "Jared events." Trump would say, "I never get any credit for this, and I'm not doing any more, it's not worth it." Then Jared would come back with just one more event, every time.

IT WILL SURPRISE NO one that after I lost the battle about hiring a new comms person, the very next day stories began to leak about the additions—or should I say addition—because only one of the people magically started to receive headlines. I hadn't yet been able to tell my teams what was happening when the first article was printed in a local newspaper. Shocking. The next weeks consisted of morning meetings between the counsel's office and my "new team." That same group along with one of my deputies was tasked with going to the Hill each day, to track the proceedings and prep the congressmen who were to go on TV each day. Jared's new strategy didn't really help, though I will say that Pam Bondi did all she could to keep me looped into things. Her legal experience was also a true asset, especially when she started to go on television.

The other new team member mostly nagged our booker to get him on television—but only the friendly shows such as *Lou Dobbs Tonight* and *Justice with Judge Jeanine*, which also happened to be the shows that the president almost always watched. I also recall that when the impeachment was finally wrapping up, I saw that Jared put him on the flight manifest for Air Force One to go to a political rally. That, after Jared had lectured senior staff in a meeting a week earlier that we were going to "keep staff to a minimum for political events because it was on the campaign's dime and they couldn't waste money." So to see a temporary employee's name on the flight manifest was incredibly rich—pardon the pun.

I called the travel office and said absolutely not. Half my staff hadn't had the opportunity to fly on Air Force One, and nothing had been cleared through me, which was how things were supposed to work. I knew this person would just run to Jared, which he did. And I knew Jared would override me, which he did. A few weeks later, Jared stopped by just to let me know that this person had been a real asset and the trip to the rally had been a "thank-you to him for all his hard work on impeachment." I simply replied, "We will have to agree to disagree on all of that," which I could see did not sit well with him. From then on our relationship was never the same, and matters got worse with the arrival of the new chief of staff.

UNTIL THE FIRST IMPEACHMENT, I didn't know much about Mark Meadows. I recall being in the Oval Office once when the president had him on speakerphone. When he told Mark I was in the room, Meadows replied, "Oh, we love Stephanie." I immediately thought that was odd because I had never met the guy that I could recall. But hey, I always loved a compliment in front of the boss.

A few days later, while I was in makeup at Fox getting ready for a TV hit, one of my staffers was in the greenroom waiting for me. Both Mark Meadows and Lindsey Graham were sitting in there waiting for their turns in the makeup chair. After my hit was over, my staffer pulled me aside and, with eyes wide and face pale, said she had overheard a conversation between the two congressmen. "When are you going to take over as chief of staff?" Graham had asked Meadows.

"I think after this impeachment is over I will take the job," Meadows had answered.

Lindsey, by the way, was purportedly a good friend of Mick Mulvaney, but it didn't surprise me to hear that he could be such a snake. His response was as loyal and supportive as you would imagine it to be in a swamp like Washington, DC: "Good, because Mick knows nothing about foreign policy. We need to get you in there." Lindsey apparently knew that Mulvaney was on his way out the door; he was just looking out for himself. Like everyone else.

While I didn't know Meadows at all, I liked Mick and his staff and thought they had been doing an incredible job under very tough circumstances. I hated the thought of their being blamed for shit that was out of their control, as had happened to the two previous chiefs. I immediately let Emma Doyle know of the conversation so she could warn Mick if and when appropriate.

THAT WAS A REALLY tense time. It was no surprise, of course, that the president was blaming Mulvaney for the problems he himself had created. Trump was railing about everyone at that time. He called Federal Reserve chairman Jerome Powell "a dirty bastard, that motherfucker" and even complained about the ethanol lobby. "There is no one more difficult than these ethanol people," he said, before adding "aside from the Palestinians." And it wasn't just Trump who seemed to be coming unglued.

In October 2019, with impeachment in the air, Democratic and Republican congressional leaders came to the White House to meet with the president, the secretary of defense, Mark Esper, who had replaced Jim Mattis, and others to discuss a path forward in northern Syria to deescalate tensions in the region. The Democrat-controlled House had just voted to condemn Trump's decision to pull troops out of northeast Syria. Nancy Pelosi, never a fan of Trump, arrived at the meeting and already seemed pissed off. Even as she arrived, she expressed annoyance at having to give up her phone, which everyone else in attendance had agreed to do as it was standard protocol for security reasons. She eventually did turn over her phone, but she didn't seem happy about it.

I sat in the room taking notes, sitting along one of the walls of the Cabinet Room with other staffers for each of the principals sitting at the table.

Now, the president did not like Nancy Pelosi, and the feeling was obviously mutual. She constantly made it clear both in public and probably in private that she thought he was borderline insane and a danger to the nation. The press would later characterize that latest meeting, which I think might have been their last, mostly from Pelosi's perspective, that Trump had "lost it" and had a "meltdown." From my seat, he wasn't the only one.

When the meeting began, Pelosi didn't waste any time trying to get Trump's goat, which was easy to do. She said, "The House has voted to oppose pulling troops from Syria." We all knew that. That's why we were there.

Trump didn't like that of course. "Congratulations," he snapped. "Obama and the Democrats gave Russia power."

At some point, Pelosi or one of the other Democrats in the room pointed out that General Mattis had publicly denounced the plan as well. "Mattis," Trump said dismissively, looking around. "World's most overrated general."

I was taking notes so fast that I wrote down that an "angry Dem guy" asked Trump, "What is the plan to deal with ISIS?" The "angry Dem guy" was the Democratic minority leader, Chuck Schumer.

Trump's response was pretty typical—brash and detail free. "The plan is to keep our country, our troops safe," he said.

Things got tense quickly. At one point, Pelosi stood up and pointed at the president. "All roads with you lead to Russia," she snapped. "You gave Russia Ukraine and Syria."

Trump looked annoyed. "You're just a third-rate politician!" he shot back.

With that, the Democratic leadership got up and walked out. I don't know if it was planned or what, but they headed straight to the microphones that were stationed outside. Well, maybe not straight there. First they stopped to retrieve their phones. That was when Pelosi seemed to blow up.

While the White House receptionist, affectionately referred to as ROTUS (receptionist of the United States), who was caught off guard by the early departures, stumbled around looking for the correct phones, Pelosi went ballistic in my opinion. "We can't even trust you people with our phones!" she yelled. "I never should have given you guys my phone! I know you bugged it!"

Pelosi then started calling the young woman incompetent and raged that there was not an umbrella available to her so she could go out to the microphones in the rain and berate the president in front of the cameras. I know that so many people hated Donald Trump, they would excuse anyone for anything if it was done in opposition to him. Still, Pelosi screaming at a young girl was totally unprofessional behavior, and it was exactly what she accused Trump of doing all the time—bullying. I will say, though, that Trump could bring out the worst in people too.

Around that time, he did the same to me.

IT WAS ONE OF the moments I had been dreading since I had taken the job as Trump's spokesperson—those requests of his that made me cringe.

That October, at the Sea Island Summit political conference, hosted by the *Washington Examiner*, General John Kelly commented on the administration he had recently departed. With talk of impeachment in the air over the Ukraine scandal, he suggested that he could have prevented the disaster. "It pains me to see what's going on," he said, "because I believe if I was still there or someone like me was there, he would not be kind of all over the place." He also said that before he had left he had warned the president not to hire a yes-man as chief of staff, among other comments that had pissed the president off.

That Saturday, the president called me. I always dreaded the weekend calls because I often didn't know what to expect or what he was seeing that could set him off. On that occasion, he had been watching the coverage of General Kelly's comments. He wasn't raging like before, but he was certainly not pleased and wanted me to "push back hard."

I was sitting on my bed, with my notebook in front of me. He talked for a while about how nothing Kelly had said was true and offered his usual comments about people who had left: that the general was totally ineffective, weak, "I fired him." Then he said, "You need to get out there. Put something out. Ready? Take this down."

That was always a bad sign. It was the way he dictated tweets to Dan Scavino as well. He proceeded to dictate a quote to me, changing it in midsentence several times, making it hard to keep things straight. He finally decided that I should say, "I worked with John Kelly, and he was totally unequipped to handle the genius of our great president."

He then had me read it back to him to make sure I had it exactly right, something I dreaded because he would get irritated if it wasn't right—and sometimes he would forget that he had made changes. But I read it back. "Good," he said, satisfied with himself. It wasn't lost on me how weird it was for someone to call himself a "genius," as Trump did on more than one occasion ("a very stable genius").

I cringed when I heard myself read it back to the president. First, it made me sound like a complete idiot, your typical crazy Trump cultist. Second, I hadn't liked everything General Kelly had done as chief of staff, but I respected him and his service. Slamming him just didn't sit right with me. As with the first lady, I was once again in the situation of striking back at someone in my own name on behalf of someone else who was actually the one who was pissed.

Anyway, I couldn't tell Trump that. Why didn't I simply refuse? That option didn't even occur to me. He was in such a bad mood. He would have flipped out. It was another weird test of loyalty, and I was determined to pass it. But I didn't like it.

I even called a few different people before I put the statement out. I wanted all of them to tell me it was as bad as I thought it was—which they all did. It took me a good twenty minutes before I found the nerve (the president would call that "weak") to respond to the press inquiries I was receiving. To be honest, I would have waited longer, but I knew the president was probably watching TV, waiting for that quote to appear verbatim on his screen.

As much as I absolutely did not want to say it and it was certainly not in my voice at all, I was the spokesperson for the president of the United States, so in my mind it was my job to say what I had been told to say, word for word: "I worked with John Kelly, and he was totally unequipped to handle the genius of our great president." Even thinking about that now, years later, I cover my eyes with shame. I'm sorry, General and Mrs. Kelly, that I didn't have the nerve to say no. It is one of my biggest regrets.

Then came the obligatory round of humiliating coverage of "my" comment on General Kelly. MSNBC said it had "a decidedly North Korean tone." A veteran columnist at my hometown paper, the *Arizona Republic*, observed, "To suggest that a man like Kelly was 'unequipped to handle the genius' of Trump is not just proof that Grisham is hooked on the president's Kool-Aid. It's proof that she's ODed."

A few days later, the president called me. He told me with some excitement about the raid that had killed Abu Bakr al-Baghdadi and then added, "I appreciated your comment about John Kelly not understanding the genius of our president." As if he didn't remember that he had composed it in the first place. I gently tried to remind him of that, but he didn't seem to care.

16

Hidden Enemy

The unseen enemy is always the most fearsome.

—GEORGE R. R. MARTIN

The year 2020 started on a promising enough note for the president, at least considering all we'd endured the previous three years. The economy was good, the impeachment effort on Ukraine looked like a dud, and the polls were decent. The press was covering so many Trump scandals and conspiracies—tool of the Russians, Charlottesville, Stormy Daniels, "shithole" countries, Sharpiegate, Ukraine, tax returns, Trump Foundation, and on and on—that I don't think people could keep track of them all.

By this time, however, I was regretting taking the job(s) in the West Wing and angry at myself for having had the ego to think I could change how things worked. I had seen sides of some of my colleagues and the president that the East Wing had shielded me from, and I didn't feel good about the way I was being perceived not just by colleagues but by the media and general public. I had started to receive hate mail at an alarming rate, even at my home, and the expectation to be "tough" or "completely loyal" to one man created a lonely feeling in me because it was never talked about, just an unspoken expectation. In fact, there was an evening around that time when I watched *Bohemian Rhapsody*, and a line in the movie really hit home. Toward the end, Freddie Mercury says, "Do you know when you've gone rotten, really rotten? Fruit flies. Dirty little fruit flies, coming to feast on what's left." The quote encapsulated how I was feeling at that point: not great about the choices I'd made, the things I'd stayed quiet about, and some of the people I was surrounding myself with.

Trump, on the other hand, was for the most part in a good mood and, as always, up for chatting about anything under the sun. Not long before, a young boy started publicly challenging Trump to go vegan in TV ads and on highway billboards. At one point, I playfully asked him if he would ever consider doing that, since the challenge was for just one month and it would raise a lot of money for a good cause. I knew he loved his steaks and cheeseburgers, but one month didn't seem that long.

Trump's response was swift, and his tone was very serious. "No, no. It messes with your body chemistry, your brain," he said, referring to something he'd apparently heard about vegetarian diets. He added, "And if I lose even one brain cell, we're fucked."

IN LATE FEBRUARY 2020, the president and first lady traveled to India. Everything about that visit was rushed, but more than that, it was planned like no other foreign travel we had done before. The president had tentatively agreed to the trip during a bilateral meeting, and it had been added to his calendar as a placeholder. But that was before a new, contagious disease called covid-19 began spreading across the world. As the date grew near most of the senior staff, and the first lady, started to have misgivings about the travel because of the new virus.

For whatever reason, Jared Kushner was insistent that we go, and as he was the "real" chief of staff, that carried weight. A final meeting was set in the Oval to determine whether the trip should move forward. Among the attendees were the president and first lady, Lindsay Reynolds, Mick Mulvaney, Emma Doyle, Robert O'Brien, Tony Ornato, and myself. As we were waiting to enter, Jared and Ivanka blew past us and into the president's private dining room to speak with him privately first—shocker. The start of the meeting actually coincided with the impeachment vote, so we all ended up watching that together before discussing the India trip. Although the Senate acquitted him, the president was in a sour mood, and made his thoughts clear to the room, saying, "I don't really want to go. It is a long trip for not even two days, and we're dealing with covid. I'll explain to Modi that it isn't a good time, and I will come later, in my second term." Jared chimed in to remind the president that with all the visits he had already promised to undertake "in his second term" he would never be in the United States to do his job. When the first lady raised her concerns about covid, many in the room assured her that the virus hadn't really hit India yet.

The president stuck to his original plan to cancel the trip. Then Jared said, "Okay, but you should talk to Modi personally to tell him." This showed just how well Jared had his father-in-law's number because, like the rest of us, he knew that the president had a hard time saying no and that Modi would likely talk him into going. Still, when we left the Oval, in our minds, the trip was off.

It turns out that Modi was somehow unreachable, and the next day the ambassador delivered the official invitation, meaning if we refused, it was a potential official snub. The rest of the planning then suddenly seemed to fall into Jared's portfolio. Everything from Modi or his top advisers came to the chief of staff's office through Jared. We had originally agreed to two cities, then suddenly Jared had us visiting three. In fact, an entire road and parking lot were being paved for our visit in the third city before we even agreed to it. All the while Jared was promising the president "huge crowds," especially in Ahmedabad, where they were erecting a giant stadium just for the visit, no matter that five days before our arrival, the stage collapsed. We immediately wanted to move to a new, safer venue, but again Jared intervened, promising that everything would be fine.

Once we actually arrived in India, Jared and Ivanka (get ready to be surprised) inserted themselves into the visit to Gandhi's ashram, something only the president, first lady, and Modi were supposed to take part in. In fact at the end, Javanka made the entire motorcade wait as they took their time getting in their own car. To this day I don't know why that trip was so important to Jared, or what, if anything, he got out of it. Jared and his team also ended up negotiating Secret Service assets and security issues that the Indian government had concerns over. It felt completely irresponsible and against protocol, which is the epitome of Jared Kushner in the Trump White House.

NO ONE IN THE Trump inner circle seemed to be taking the new virus too seriously at first. During a meeting with Modi in India, Trump mentioned the thirty-four people who were suffering from covid-19 in quarantine on military ships. He complained that the news was affecting the stock market. "I wonder if this is overrated versus the flu," he said.

Of course, those thirty-four people would not be the only ones to contract the disease. As the number grew and grew, Trump still seemed resistant to doing anything too drastic. Contrary to what he would say later, he didn't immediately want to ban travel to China. And he asked officials in the White House if we were making "too big a deal out of this."

In the last week of February, Italy saw a major uptick in cases, covid-19 surfaced in Brazil (which would go on to have a major impact on the White House), and the United States reported its first death, in Washington state.

For me, personally and professionally, things started to truly go to shit the day that Mick Mulvaney resigned. This is by no means trying to downplay or marginalize the virus that was about to sweep across our nation and affect all of our lives for the next year and more. Mick announced his decision on Friday, March 6, 2020. We had been traveling all day. We left the White House that morning, headed to Centers for Disease Control and Prevention (CDC) headquarters in Atlanta, Georgia, to get an update on covid-19. The traveling crew was small and included myself, Emma Doyle, Michael Williams, who also worked in Mick Mulvaney's office, Jared Kushner, Dan Scavino, and me. Mick was away, and Emma was his representative for the trip. She and I had grown close, and I noticed that she was not her usual positive and uber-professional self that day. She seemed a little bit down.

Being the chief of staff rep on a presidential trip is no joke. That is the person who must keep the government running if serious shit hits the fan. After the CDC visit, during which the president had argued with reporters about the government's response to the pandemic, we were all a bit tired and grumpy and frankly ready to go to Mar-a-Lago. We were sitting alone in the senior staff cabin when Emma finally told me that on Marine One that morning, the president had looked at Jared and asked, "Should we tell Emma what's going on?"

When Jared hesitated, because he never trusted anyone and loved being the keeper of all information, the president prodded him, saying "You can trust her, and besides, maybe she'll stay."

Jared then told her that the decision had been made that Mick would leave but the president had not yet made up his mind on a replacement, leading Emma to believe that there was still time for him to change his mind. Nevertheless, Emma was devastated by the news. She had worked with Mick in his congressional office and at the Office of Management and Budget prior to their moving to the White House, and thus they had a close-knit bond.

The president and Jared then told her to keep the news quiet because no one else knew yet, something I could verify because even though I was the press secretary, it was the first I was hearing of it. Emma was understandably upset but glad that they'd given her the heads-up and time to process what it meant for her team, and for her. I spent the rest of our flight doing all I could to make her laugh, and we made plans to eat at our favorite restaurant in West Palm Beach, Kapow! Noodle Bar, and try to get in some shopping or beach time if we could. We landed in warm Palm Beach that evening and, as per usual, loaded into the motorcade for the ten-minute drive to Mar-a-Lago.

That weekend neither Emma nor I was able to stay on the property because the club was so full, and when that happened only the body man and Dan Scavino were given rooms at the resort. Believe me, if Scavino had ever been told he needed to stay off property, we would all have heard about it the entire weekend. He had become another one with an outsized sense of importance about himself—and I'm including myself on that list, by the way. Though Stephen Miller also became something of a diva on trips—demanding special rules for when his bags would be picked up in the morning, for example—nobody was quite like Scavino. He insisted on being in the staff van closest to the president's and threw a fit when he wasn't. He wanted to be on the same floor as Trump in hotels and threatened to quit on more than one occasion when he felt, as he put it, "I'm fucking being disrespected." "Who the fuck do these people think they are?" By the end of the president's term, it got to the point that we tried to cut him off at the pass. If I had a room or seat he wanted, I gave it up to him ahead of time. So did others. He was being a baby, for sure, but he was pretty crucial to the president. And let's be honest, if he left, the rest of us would have to deal with Trump a whole lot more.

In fact, as the chief of staff's representative, Emma was supposed to stay on the property in case anything major happened, but fighting Scavino was just not worth it. As we pulled into the driveway of Mar-a-Lago to wait for everyone to unload, I got a call from Scavino. He wanted to give me a heads-up that a tweet would be going out "in a few minutes" announcing that Mick Mulvaney would no longer serve as chief of staff. I asked the natural questions that any press secretary would need to know: When was it effective, was it amicable, who would be replacing him? Scavino didn't have any of that information. They knew but just didn't care that seconds after the tweet went out, it was my phone that would start blowing up and in the absence of answers, the speculation would begin. Looking back now, I think the president enjoyed the turmoil and the frenzied speculation—no matter how chaotic and uncertain it was for his own staff and, I don't know, the country, even the world, during the start of a global pandemic. Couldn't we keep the lid on the drama in the Trump administration for one damn day?

Making it worse, I was in the staff van with Emma and Michael Williams, and as I was about to tell them what was going on, the tweet went out (so much for the "in a few minutes" time frame I had been promised). To put it mildly, the fifteen-minute ride to our hotel sucked. Emma was crying, Michael was visibly upset, and I was angry that friends and colleagues I respected were hurting. I had grown to really like Mick and his entire team. They were inclusive, fun, whip smart, and able to deal with the huge cast of crazy characters in our administration.

When we arrived at the hotel, we agreed to meet for a drink in the lobby bar, and Dr. Sean Conley, the president's physician, said he'd join us. I spent the next twenty minutes in my room telling the press that I had no further details about Mick's departure or replacement. Of course, most of them didn't believe me, because they rightly assumed that a communications director/press secretary would know what was going on in the White House. That only compounded my own frustration about how little I knew on most days. The rest of the weekend, we were all in our own rooms, frustrated, confused, annoyed, uncertain about what was going to happen next. I had a lot of work to do (mostly saying "I don't know anything" to one person after the other), and the rest of the crew understandably wanted to lie low. In fact, Emma and I had a conversation about how she planned to skip the meeting with the Brazilians and the fundraising roundtable that was scheduled. She said she didn't need people staring at her with pity or asking questions, which I completely understood. I actually planned to stay away myself but then felt guilty because Brazilian president Jair Bolsonaro's press secretary, my counterpart, would be there. That was a big mistake, as I would soon find out.

ON SATURDAY AFTERNOON, I headed to Mar-a-Lago to meet with my Brazilian counterpart. The meeting was uneventful, more of a courtesy meet and greet. We sat on the couch in the ornate large room referred to as the "living room" in the main house for about an hour and exchanged pleasantries. Soon after the visit Bolsonaro's press secretary would announce that he had tested positive for covid-19.

A few days after our meeting, I had the sniffles. But I had no idea if I was psyching myself out or what the deal was. Covid was still so new to everyone at that stage, and there was still much to learn about it, including means of transmission, incubation time, symptoms, how long to quarantine, and other important matters. But as I worked for one of the biggest germophobes around, I immediately told Dr. Conley about my Mar-a-Lago visit. He advised me to go home to be safe, and that I would need to go to Walter Reed National Military Medical Center to be tested. Frankly, I was more than happy to go home. The boss had been in a bad mood, I didn't feel great, and my principal deputy, Hogan Gidley, was always thrilled to have face time with the president—so in my mind, everyone was winning.

The next day, I headed to Walter Reed to get tested. When I saw the very long, very frightening stick one of the doctors was pulling out of a clear plastic bag, I said flatly, "You are not putting that up my nose—right?" The other doctor in the room said, "Yes, and we need to do it twice." I then asked exactly how far it would have to go, and when the doctor holding the torture device showed me the mark at the very bottom of the stick, I backed up, grabbed the sink behind me, and said, "No, nope, no way that is going up my nose."

Over the next fifteen minutes, I probably said "Fuck you," "Fuck off," "Fuck, no," "This is bullshit," and "I truly fear I will punch you" a few times each. When he finally got the swab up there, it was just as bad as I'd thought it would be. Not only does it feel like it's poking you in the brain, the doctor swirls it around for good measure. And he did it two times, no less. Anyway, I made it through the process alive, apologized profusely for my behavior, and headed to my apartment to wait for the results. On the way home I received a text message from Dr. Conley saying "I hear you cursed like a sailor, I expected nothing less." The man knew me well.

BY MARCH 11, 2020, a Wednesday, I had a slight cough and sore throat. It was a well-known fact that if you had any kind of cough or cold symptoms, you were supposed to steer clear of the president, so I did my best to avoid him.

Around noon that day, one of my deputies came into my office to let me know that they had stumbled into a meeting among Hope Hicks, Jared Kushner, and Pat Cipollone. The three of them had apparently been discussing the need for the president to give an address to the nation on covid-19 from the Oval Office that evening.

We had nothing new to announce, but that didn't seem to matter to them, and I had been around long enough to know that there was no use trying to sway that particular group. Adding to that challenge, the Office of the Vice President was to head up all covid communications, as Pence had been named head of the recently established White House Coronavirus Task Force. So on one of the most important issues facing the administration—and the country—I took a back seat.

An address to the nation is serious stuff, and whenever possible you need plenty of time to prepare properly—unless, of course, you were in the Trump White House, where everything was like a clown car on fire running at full speed into a warehouse full of fireworks.

A couple of hours later, a meeting was called in the Oval Office so that members of the Coronavirus Task Force could brief the president on the latest involving the virus. I wasn't invited, which was typical. Meetings just "happened" all the time in that White House. Random people would wander into the Oval Office and start talking about random things, and suddenly something would be decided or Trump would agree to do something—and anyone who wasn't in the room would find out about it later on Twitter or on cable news. I can't even count how many times Trump would announce something and I'd say to someone, "WTF?" and the reply would be "Oh, yeah, there was a meeting . . ." If you were on the comms staff, the vice president's staff, or the White House counsel's team or, God forbid, were a member of the cabinet nobody ever thought about, and you didn't happen to know about some meeting that had happened spontaneously, well, that was your problem. Put another way—imagine being a subject matter expert in the Trump administration and there was a meeting on your subject, something you knew more about than anyone else, something you'd slaved away for months on to come up with a coherent workable plan, and no one told you the meeting was happening while decisions were made by people with a tenth of your expertise on the subject. Well, welcome to the Trump White House. That's how we rolled.

But that time, I caught wind of the covid meeting when I saw it had been added to the president's schedule. I suspected that Jared or the recently returned Hope Hicks had called it and determined who was to be invited. I had learned by now just to walk into Oval Office meetings whether I was invited or not. Even with my cough, I thought that the meeting was important enough to attend without an invite and with throat lozenges galore.

In attendance at the meeting were the two new stars of the Trump administration, Dr. Anthony Fauci and Dr. Deborah Birx. I guess you could include Robert Redfield, the CDC director, on that list, but he was kind of an afterthought. Fauci and Birx—especially Fauci—ran that particular show.

Trump had liked Dr. Fauci—for about ten minutes. Then he had decided, as most everyone in the White House did, that Fauci was a showboat who liked seeing his face on television. He hated the way Fauci always talked about worst-case scenarios and thought his statements scared people, hurting the economy and Trump's reelection chances. Fauci would say, of course, that he was just telling the truth as he saw it. But he was also exceptionally savvy at handling the media, and they adored him. The Office of the Vice President tried to keep him in line and make sure his statements were coordinated with ours, but it seemed that Fauci couldn't care less what we thought. He had the power of the media behind him. I will say this: he sure knew a heck of a lot more about covid and other infectious diseases than the rest of us ever could. So I couldn't blame people for listening to him. But let's not pretend he didn't love being a media hero.

If Fauci was a master at playing the outside game with the press, Birx was better at the inside game. She knew how to tell the president what he wanted to hear or at least give him news in the way he preferred to hear it.

The meeting was packed. Dr. Redfield, Dr. Birx, and Dr. Fauci were sitting in front of the Resolute Desk, along with Vice President Pence and Treasury Secretary Mnuchin. Marc Short, the vice president's chief of staff, was standing behind him. I sat on one of the couches with Ivanka to my left, which I found odd as she hadn't been involved in anything to do with covid until now. Jared Kushner stood behind us, and National Security Advisor Robert O'Brien (who had replaced John Bolton) and his deputy, Matt Pottinger, were sitting in the chairs to my right. Across from me on the couch was Keith Kellogg, Pence's national security advisor, and standing behind him was Hope Hicks.

Hope had left the White House about two years earlier. I think she had been understandably stressed out by the job and didn't like the fact that she had been subpoenaed and had had to testify in the Robert Mueller investigation (during which it had been revealed that she had told some "tiny white lies" on behalf of the president). She was also being hounded by the paparazzi about her personal life. She had left for a sabbatical at Fox, where she had a great title and reportedly got a seven-figure salary.

I can't pretend that her presence didn't irritate me. In my eyes and the eyes of others who had stayed to deal with all of the craziness, Hope had taken the easy way out. We all would have loved to take a cushy job somewhere else for two years so that we would be begged to come back to the White House to "save" the administration. Now she was meeting with the president one-on-one to give advice about communications strategy—and I would have resented anyone doing that, not that Hope cared, which again I strangely admired.

By the way, that was March 2020. No one was wearing a mask—not the president, not Fauci, not Birx. There was no social distancing. The subject didn't even come up.

The meeting began in a pretty standard way. Members of the task force updated the president on what information they had, the numbers of infected by country, and the projections for the weeks ahead, which were quite sobering. Their recommendation was to temporarily close the country's borders to travelers coming from Europe. Obviously that was seen as a drastic move, the logistics of which would be huge and complex. If there were US citizens across the pond, could we get them home first? What about connecting flights from other countries to the United States? How would the move impact the economy? What about trade? How long would the travel ban last?

In the middle of all the discussion, Ivanka kept chiming in, "But I think there should probably be an address to the nation tonight."

I let that pass because in my mind there was no way we could pull one off with no speech prepared, no communications strategy, no consensus, and only a few hours' notice. We did a lot of random things, but that just seemed too crazy even for us. Would I ever learn? As the discussion continued, Secretary Mnuchin kept raising the potential impact on the economy. He felt that the recommendation to shut down the borders was far too severe and the financial impact to our country and the world would be something we would not recover from for years.

The discussion got quite heated, especially between the secretary and National Security Advisor O'Brien, who at one point said to Mnuchin, "You are going to be the reason this pandemic never goes away."

Hope Hicks continued to chime in with questions and ideas that had been discussed weeks before. And Ivanka, the women's rights, small-business, covid, crisis communications expert, just kept repeating "There should be an address from the Oval." Finally, the Princess turned to her most powerful ally besides her father. "Jared, don't you agree?"

Any guesses as to what Jared replied?

At one point I called her out on it. "What is it we'd be saying?" I asked. Because if she had a message she wanted her father to deliver, it was still a mystery to me.

She just looked at me, seemingly confused as to why I didn't think it was a brilliant idea. Birx, Fauci, and the other professionals in the room just watched all the nonsense without comment. To their credit, they pretty much kept straight faces, although I imagine they thought they were surrounded by lunatics.

In my mind I kept saying "This is not a reality TV show. We cannot address the nation with a bunch of mumbo jumbo just so he looks presidential. That's not how this works." This was some serious shit, and all they were thinking about was TV and image and optics. But as I say, that was just in my head. One of my other biggest personal regrets is that I didn't have the courage to speak out against Jared, Ivanka, and Hope about the plan and what a disservice it was to the country and the president. But by now, I knew it was fruitless; they would do what they wanted to. People were afraid to stand up to them and I have to admit I was one of them. Trump let that state of affairs go on all the time.

But I should have at least made my thoughts to the president clear; I could have found some moment to say my piece. In fact,

I was impressed with Secretary Mnuchin for that very reason. He kept pushing back, over and over, against a roomful of people who supported closing the borders completely. I did not agree with his point of view, but I admired his willingness to speak his mind to the group.

After about an hour of going around in circles, the president told us all to go to the Cabinet Room and "figure out what to do."

I remember thinking to myself how ridiculous it was that the president of the United States had to tell his own staff to go figure this out and then come back to him. To his credit, he told us that it didn't matter how long we took but to come back with a plan. It occurred to me that when the meeting had been called, we should have come in with a recommended plan. We were doing things backward.

We all headed to the Cabinet Room. I sat at one end of the long table next to Marc Short and the vice president's press secretary, Katie Miller, who had walked in. Across the table was the vice president with Jared to his right. Directly across from them were Mnuchin, O'Brien, and Matt Pottinger. At the other end of the table, Hope and Ivanka huddled. The medical members of the task force were sprinkled around the table. We were coming up on 3:00 p.m., and I kept thinking it wasn't going to be good. It started to seem inevitable that an address to the nation from the Oval Office was going to happen that night—and we had no idea what the president should say.

The discussion began, and it was much like the previous one. Most everyone except Secretary Mnuchin agreed that we needed to close the borders to flights from Europe. What struck me in that meeting was that Jared, who was sitting next to the vice president of the United States, commandeered the meeting and was calling all the shots. As many times as I had seen him behave that way with members of senior staff, that particular time made me uneasy because it was the vice president. It was disrespectful, and I remember feeling both embarrassed and disgusted. Ivanka was also doing her "my father" this and "my father" that routine, making it impossible for staff members to argue a contrary view. At some point I think Dr. Birx decided she'd ridden on the crazy train long enough and excused herself to get back to work.

I used that opportunity to leave as well. I knew I needed to get my team together so we could begin work on whatever it was that we'd be announcing, and we had to call the networks ASAP to see if they would give us the airtime for an address. By that point, we were giving them only four hours' notice, and I wasn't sure they would even agree. When networks give up program airtime, they also give up advertising, which is how they make money.

I gathered my team, told them that the president would be addressing the nation shortly, that details were still being worked out, but that we needed to get a statement, press release, and fact sheet together. I instructed one of my deputies to call the networks to reserve airtime for 8:00 p.m.—which no one else had even thought to do.

Katie Miller did her best to keep us looped in as to what the announcement would be, and we waited. Luckily, Katie is married to the president's speechwriter, Stephen Miller, so she went to his office and sat there while Jared Kushner frantically dictated to Stephen, who wrote something out, sending me updates as she knew them.

Meanwhile, members of the press, having caught wind of the address, lined up outside my door to find out what was in the speech. I wish I knew! They were as frustrated with me as I was with myself. Unable to do the basics of my job, I felt helpless and demoralized. And the more I thought about it, the more outraged I grew at Jared's behavior. He was not an expert on any of those things—shutting down borders, the economic consequences, the health consequences—yet he alone seemed to be deciding our first actions to address one of the most devastating crises in our nation's history.

I had shared with Mrs. Trump many times my opinion that if we lost reelection in 2020 it would be because of Jared. She didn't disagree with me. It was my fervent opinion that his arrogance and presumption had grown over the years, and he threw his power about with absolutely no shame. I would venture to say that being in the White House changed Jared as a person. There was no reason that he should be sitting with the speechwriter laying out our nation's plan to fight a global pandemic. And I knew that if things went badly with the speech, which felt inevitable, he would be the first person to say in the president's ear that the comms team had fucked it all up. He was Rasputin in a slim-fitting suit.

For the next couple of hours, things were frantic in the West Wing, which is saying a lot. The networks agreed to the airtime, so we had to get their crews in the Oval Office to set up lighting, cameras, and microphones. My team and I were writing furiously, based on what little we knew about the speech and some information Dr. Birx had provided us. We were also doing our best to preview the remarks to some of the major outlets. We were now thirty minutes away from what might be the most important speech of the Trump presidency, a speech that could determine his reelection, and it was still being written. There was no time for fact-checking, vetting, or notifying friends and allies on the Hill or abroad. There was hardly any time for the president to read it and make changes to it. It was a total clusterfuck from start to finish because Ivanka and her crew wanted her father to be on TV.

Finally it was time, and the president addressed the nation from the Resolute Desk in the Oval Office. I remained in my office to watch. Like almost everyone else in the administration, I was hearing much of the speech for the first time, along with the rest of the country. Some things about it were good. His tone was different than usual—more serious and reassuring. And the speech did make it sound as though he was doing something. I just wish we'd had a chance to think it through better. Maybe, I hoped, we'd dodged a bullet. But as the speech went on, my inbox began filling up with messages and questions from reporters. That wasn't good. The speech contained a number of misstatements and sloppy wording—some caused by the president stumbling over a few phrases—that sowed confusion about such things as which countries would be affected by the travel restrictions and if international trade would be banned. News outlets all over the world picked up on the discrepancies in the speech. People from various federal agencies started to call and ask us how to explain or clean up some of the things that had been said. Once again a line of reporters formed outside my office. Of course, it was our problem, not Jared's or Ivanka's or Hope's. No, they were in the dining room off of the Oval Office, Trump's usual hangout, congratulating themselves and telling the president how awesome he was. When the press reaction to the speech turned bad, I assumed all of them would be first in line to blame me or the comms shop for botching it all up. But as it turned out, I would have other problems to deal with soon.

As mentioned earlier, the Brazilian government announced that the president's press secretary, whom I'd spent time with, had tested positive for covid. Because my symptoms started to get worse, I was put into quarantine away from the White House, which set off a chain of events over the next three weeks that would change my life forever.

17

Chief Four

It's so nice when toxic people stop talking to
you. It's like the trash took itself out.

—UNKNOWN

I quarantined in my apartment and got much sicker over the next week. Eventually, I didn't have a voice and slept many hours of the day away.

In the meantime, Mark Meadows had begun his new role as chief of staff in earnest. After Priebus, after Kelly, after Mulvaney, it finally dawned on me what was going on—it only took me three years. Trump, Ivanka, and Jared were the ones calling most of the shots in the White House, but they wouldn't blame themselves when things were fucked up. No, the fault for plummeting poll numbers or bad press or confusing policies or the inability to keep some promise the president had made or some scandal always fell to the staff. And the solution was always to get rid of the person, aka the scapegoat, who was letting them down in favor of some new, temporarily perfect replacement—who for a few weeks would be a modern version of a James Baker or a Roy Cohn or a P. T. Barnum. I guess that somewhere in the back of my mind I knew that after Mulvaney left, I'd be scapegoat number two—another reason Sarah Huckabee Sanders had been smart to quit before that fate befell her. Maybe it was also the reason Hope Hicks had taken her leisurely sabbatical—so Trump would miss her and she could escape being a scapegoat, too.

Anyway, now that I was in quarantine, the new golden boy, Mark Meadows, started meeting with my staff and requesting calls with me to ask about different members of my team and why I had structured things the way I had. Now, pay attention, because this is probably the one and only time you will hear me say something sympathetic about Mark Meadows, who in my opinion may be one of the worst people ever to enter the Trump White House: he came into the administration at an exceptionally chaotic time, and that is saying something. The covid-19 situation was evolving every day, his press secretary and communications director was the first person in the administration out potentially with the virus, he was still getting to know the players inside the building, and he was no longer just being the guy who talked to POTUS on the phone at night to shoot the shit. Now he had to take hold of this mess.

If there was a scale of awfulness in the Trump White House, with five being the most terrible person around, I'd give Mark Meadows a twelve. I know this will sound as though I'm bitter because of the role he'd eventually play in my departure as White House press secretary—and fair enough, because I was pissed as hell—but I stand by that assessment completely.

First, a little backstory so you know it's not just me. In his memoir, former speaker of the House John Boehner recounted how Mark Meadows, when he was a newly elected member of Congress, had fallen to his knees on the floor of the speaker's office and begged Boehner's forgiveness after he had voted against Boehner for speaker. Then, after Boehner forgave him, Meadows voted against him later anyway. So . . . that story makes Meadows look like a phony (check) who tells people what they want to hear (check), kisses the ass of whoever is the boss (check), and will screw you again anyway if it serves his purposes (check again). With his syrupy sweet southern accent and "Aw shucks," "I love the Lord" demeanor, he could be Andy Griffith—if Andy Griffith had an evil twin. To me, if he were a movie character, you would hear thunderclaps in the sky and menacing organ music whenever he entered a room. In summary, I did not like Mark Meadows.

I assume Meadows had promised the president that he would get him good press and find all of the leaks, and in return I'm willing to bet all I've got that the president had told him that he could run things however he wanted, hire anyone, fire anyone, and it was "his show." Of course, POTUS had said all the same things to Reince, General Kelly, and Mick. With those two missions in mind—get good press, and find the leaks—Meadows went to work.

As I observed him, I noticed a familiar pattern to how he operated. He would call people into his office, tell them what a great job they were doing, how much the president liked them, and then cut them out. He stopped people from hiring anyone he didn't approve of. He exiled people he didn't know to the Siberia of the Executive Office Building and then pretended he was doing what he could to bring them back. It was sadistic mind fucks 24/7.

I saw him do that to my friend Emma Doyle, who had worked so tirelessly and effectively for Mulvaney. Emma had offered to brief Meadows on the chief of staff role and what he needed to know. He'd said thanks, sure thing, great idea, and never called her in. He told Emma he wanted her to stay, then suddenly moved her across the street. He reached out to one of my assistants, who had also been evicted to the netherlands, called him into his office, and said, "Tell me what job you want here." So my assistant suggested a job, and Meadows said, "No, you can't have that one." My assistant then suggested something else. Meadows shook his head. "Oh, no, that won't work either." On and on. He also put a stop to every pay raise and title change and promotion I had lined up for members of my staff but didn't bother to tell any of them.

At one point, he even dared to get into Jared's business, telling Jared that he couldn't bring in people from other agencies without his approval. Jared told him, politely I'm sure, to go fuck himself. Meadows was smart enough to figure that one out right quick.

It was suspected that Meadows leaked against people in the White House to the press quite a bit. I also watched from afar as he took full advantage of my quarantine to get all into my operations. Prior to my quarantine, I had sat down with him to go over the structure of the teams, who people were, and what they did. He did not seem to care much about any of it and instead informed me that he wanted to bring briefings back and that he had some people he wanted me to bring on board. He said that the president really liked Kayleigh McEnany on television and he wanted me to hire another person for communications. I told him I had no problem hiring Kayleigh and thought she'd be a great deputy press secretary we could put on TV as often as we could.

But when it came to the other person, I told him that was a hard "no." She had previously worked in the Office of the Vice President and had left a very negative impression on many of my team. I had never met her, but I had heard plenty from my staff and even members of the press about how she operated. I told Meadows I believed her hire would be bad for morale after just getting over having to hire someone just for impeachment. Mark told me in no uncertain terms that she had worked for him on the Hill, had done a good job, and I was not going to have a choice in the matter. I wanted to explain to him that placing a good story in *Guns & Ammo* magazine was not on the same level as being a White House communications person, but I could tell that that was not something he cared to hear. Mark Meadows, for all his syrupy southern "kindness" and his fatherly communication style, did not like to be challenged in any way. His face turned bright red, and he gave me a pissed-off look. I bet he's a terrible card player; he has no poker face whatsoever.

Jessica Ditto, my communications deputy, had already tendered her resignation. She had committed the cardinal sin of giving Meadows her opinion on covid messaging while I was in quarantine, and he was having none of it. He let me know how unhappy he was with her "attitude" and that I would need to "deal with it" when I returned. It didn't matter to him that she was wildly talented at communications and had been there from the beginning. I was actually proud when Jessica took that stand against him.

MY COVID QUARANTINE LASTED almost three weeks. I caught up on reality TV, kept up with reporters, and listened in on meetings via telephone. The president called me one day to see how I was doing. "You sound like shit," he said, but in an affectionate, joking kind of way. He was very kind on the call, telling me to stay home and get healthy.

I was legitimately sick for a couple of weeks. It was not easy to be isolated as I was, and I told everyone I spoke to that I was worried about the way the pandemic would affect people's mental health. Granted, I had the stress of a new chief of staff wanting to reorganize a team I had worked hard to build up with a big focus on morale, as well as members of my team calling me daily, upset with how things were going.

In addition, I had to take another covid test since somehow my first one had been lost (so that was now four horrific pokes into my brain for anyone keeping count). By that time, there were so many tests "in the queue" to be analyzed that after week three, when I was feeling better but we still had no results, I was allowed to go back to work as long as I stayed in my office. Once I was back, we were very strict about people sitting far away from me, and I left my desk only to use the bathroom.

NOT LONG AFTER MY return I was summoned to the Oval Office. I was feeling better and hadn't been in contact with anyone in almost a month, so I headed in. The president was in his private dining room with Jared, Scavino, and Pat Cipollone. As I approached and POTUS saw me, he held up his hand. Looking at me as though I were Typhoid Mary, he forbade me to come any further than the doorway.

"Hi, honey, how do you feel?" he asked, which immediately set me at ease despite being barred from the room.

I told him I was feeling much better, and then we got to work. I honestly don't remember what the meeting was about, except that I remember standing awkwardly at a distance from everyone—a literal outcast. At some point, maybe ten minutes in, his mood turned and he went into a rage. He started yelling at Pat Cipollone, telling him that all of his lawyers were "weak," that no one would fight for him. The usual.

In the middle of his rant he turned to me. "And where the fuck have you been?" he snapped, as if he remembered nothing of our conversation only minutes earlier. "I'm the only president with a PR person who is never around! I haven't known who to call for anything, just left with nothing!" I remembered what he had told me right after I'd been hired for the job: no one else mattered. "Only me."

He went on for a few more minutes of general ranting about how no one who worked for him was any good, he needed fighters, and we were failures. When I knew he was done with me, I left. I went into my office and cried, which, as I've mentioned, is a rarity for me. But at that point I still felt like shit physically, I was growing to really dislike Mark Meadows, and now somehow Trump had made me feel guilty for being sick and out of the office. Nothing mattered but Donald Trump—when was I ever going to understand that?

18

Headed Back East

You know you have made the right decision
when there is peace in your heart.

—UNKNOWN

Jared Kushner wasn't happy. Those were words you never wanted to hear in the Trump White House. During my quarantine, I heard from members of my staff that the president's son-in-law was dissatisfied with the White House's covid messaging and was, predictably, holding me responsible. In fact, an article appeared with an anonymous quote calling me a "nonentity." Didn't I tell you that was going to happen? Jared, who had single-handedly managed the president's covid response in the address to the nation that his wife had demanded, keeping me and everyone else out of it, was now complaining about the way our covid response was being received. Jared, who had appointed himself the expert on every problem—from the border wall to trade policy to an unprecedented global pandemic—suddenly claimed that he'd had nothing to do with the mess that was made. I cannot believe I ever found the guy attractive, sincere, or kind.

By that point, I'd had more than my fill of Jared and his stupid sayings. Did I mention his sayings earlier? Well, the first was "POTUS makes the waves. We just ride them." Whatever the fuck that meant, though as I think about it I guess it was kind of true. The second was something everyone in the White House heard all the time and made fun of repeatedly behind his back: "What I've learned from my time in government . . ." Come on, guy, you'd been "in government" for about twelve seconds. And it wasn't clear that he'd actually learned anything—except how to avoid blame and find new suckers to carry his tune until he was done with them. The latest sucker was Mark Meadows.

SHORTLY AFTER MY RETURN, Meadows let me know that some changes were going to take place to my teams. He said he would eventually replace me as press secretary but wanted me to remain in the West Wing as communications director and continue to oversee the operation. To add a little more insult to his plan, he let me know that getting a new press secretary could "wait a bit." Gee, thanks for the crumbs, guy. In the meantime, he told me to offer the job of deputy press secretary to Kayleigh McEnany. Like a good soldier, I went directly to my office and did so. Kayleigh, by the way, turned the offer down, concerned that the title would be a "demotion" for her. I was disappointed by that line of thinking, because in my opinion it is an honor to work in the White House no matter what your title.

A few days later, I received an inquiry from a reporter with Axios, someone who was notorious for how well sourced he was in Trump World. He said that he was preparing to write that per Meadows, my two roles were going to be replaced by Kayleigh McEnany and Alyssa Farah. This was not a reporter who often got his facts wrong or had unreliable sources. So my instincts told me that that was coming from on high.

I immediately forwarded the email to Mark, writing "I think it is very important you go on the record denying this, for the sake of morale and to put an end to any speculation." I sent a statement to him for approval that would come from him, reading "This is false and there are no personnel changes planned at this time." He wrote back, "No response is needed here." There was no reassurance that the story was untrue, no effort to tell me that he stood behind what he had told me earlier—that I wouldn't be replaced. He wasn't even interested in killing the story to protect the president from another round of "staff in chaos" articles in the press. Since I'd seen him fire and reassign people left and right after assuring them that they were doing a great job, I had no doubt he was trying to do the same to me. And he didn't seem to have the guts to tell me directly that I was gone.

I was mad as hell and quickly turned to the one person in the White House I could (usually) count on. Mrs. Trump read the email exchange, and we quickly got on the phone. In one of the first times since we'd met, she got a taste of my temper as I unleashed an expletive-laden rant about Meadows, his team, and even the president, who I assumed was supportive of this. That was how Trump always got rid of people—behind their back with leaks to the press or on Twitter.

Mrs. Trump offered to call Meadows, but I didn't believe it would do any good and I thought that made me look weak. She didn't seem to give a shit what he was doing to other staffers in the West Wing. But she did seem to care about his messing around with people who worked for her. She also kept reassuring me, saying "POTUS loves you. He has not mentioned this. I do not think he knows." I loved her message but wasn't sure I believed her.

I told her I wanted to resign first, and I asked her for her advice.

"What if you just come back to my team for the communications?" she asked. As tempting as that was, I explained that it wouldn't serve her well if the story was that I'd failed in the West Wing and had to walk across the White House with my tail between my legs.

Her next idea was simple—and shocking. She said, "You send the reporter a statement and say it is not true." In other words, call their bluff.

If I did that, I would directly be defying the new and empowered White House chief of staff, who had specifically told me not to respond. That sort of thing never happened in a usual White House, but of course, we weren't dealing with the usual. We often flouted the rules anyway.

As I hemmed and hawed, Mrs. Trump kept pushing. "You must be strong," she said, "and if it is not true, then what is wrong with a statement?" She added, "You need to fight for yourself, because this is not right and what he is doing is not right."

I was mad. I was planning to resign. I saw the handwriting on the wall. But I'd been loyal all those years and thought I deserved better treatment. So I decided, why not? If I'm going down, I need to do it fighting. I told Mrs. Trump I would write something up and send it to her for her thoughts. For the next few minutes we went back and forth on what to say, finally settling on "Sounds like more palace intrigue to me, but I've been in quarantine. If true, how ironic that the press secretary would hear about being replaced in the press." When the story was posted with my quote prominently in it, Mrs. Trump texted me a few of her signature emojis: a smiley face, the flexing arm, fire, and fireworks.

The next day, a Saturday, I received an email from Meadows. It read, "Stephanie, I have decided to move forward with the changes to the press shop that we discussed. While I would like you to stay on as director of communications, please be prepared to announce a new press secretary on Monday." He didn't bother to tell me who the replacement would be, by the way.

I had seen enough of Meadows's leadership style to know that I was not going to be able to protect my team or succeed in the position of communications director with the people he was forcing on me and whom I didn't believe in or trust. From a completely ego-driven side, I also assumed the press would say that I was no longer the press secretary because I wouldn't give briefings, and I just didn't want to live with that narrative (spoiler alert, I'm living with that narrative). True or not, the press would likely pit the new press secretary and me against each other, and that would be good for no one.

Once again, I forwarded the exchange to Mrs. Trump and told her I'd give her a call later as I needed time to think.

By the time I called her later, I had made up my mind again: it was time to resign. When I told her of my decision, she responded with a solution that surprised me, saying "You come back to the East Wing and be my Chief of Staff." I was so honored and grateful that she was making me the offer so that I would not have to leave. There was only one problem: she still had a chief of staff, the same one who had been there since day one.

Lindsay Reynolds and I had been extremely close for the first two years of the administration. We were inseparable and had managed to get ourselves into several hilarious and oftentimes embarrassing situations—but it made things fun. Some of my best memories are with her. Yes, there were occasional tensions between us. And months earlier, we'd had a falling-out over who should replace the outgoing deputy chief of staff for operations. Only recently had we begun to reconnect over our shared frustrations with the Meadows crew. But Lindsay had been having issues with the staff of late—both East Wing and West Wing. She had also been spending much of her time with her kids in her home out of state since schools were canceling classes with covid so Mrs. Trump wanted her to be able to continue with that. It seemed like my situation with Meadows was perfect in a couple of different ways that could benefit everyone.

Still, I was surprised by how quickly and coldly she seemed to be willing to drop Lindsay. The fact that it was now the second time I had seen her penchant for just dismissing people so quickly should have been a warning sign that she was not so different from her husband. Instead, as with Stephanie Wolkoff, I convinced myself that for many reasons the change was merited. It never occurred to me that Mrs. Trump would ever cut me off like that. My ego was too big; I assumed that I was different, too important. No one would do something like that to me.

I asked how her chief of almost three years would be told, and the first lady simply said, "Lindsay wants more time with her children. Marcia will take care of it," referring to her senior adviser on the team.

In the meantime, it gave me great joy to compose an email to Meadows saying in part "Thank you for offering me half of my job, but I have accepted the position of Chief of Staff for the First Lady" and would be heading back to the East Wing. His response to me was a little psycho in my opinion: "Bless your heart. We'll talk about this more on Monday." "Bless your heart," by the way, is known as a nice way to say "Fuck you" in the South.

Eventually Meadows called Lindsay and offered her the opportunity to resign. I was told that she was pretty shocked by how fast things were moving. There was an irony that at the time was lost on me. We were doing to Lindsay what Meadows had just done to me: she was asked to resign without warning, even though she'd worked with and been close to the first lady for years. And just as I had been with Meadows when it happened to me, Lindsay was pissed.

I would hear reports from people that she spent the next months calling many of her old (and my new) staff incessantly. She felt I had set her up and told Mrs. Trump that people in the East Wing were unhappy and that I wasn't doing a good job. Then the same staff had to tell FLOTUS that that wasn't true. Lindsay didn't deal with what happened well at all—and I understand that. Her anger and blame were misplaced, but it's not as though she was going to turn her ire on Mrs. Trump. Instead she talked herself into thinking I had manipulated Mrs. Trump into forcing her to resign—even though she knew full well that no one manipulates that woman.

IT WAS AGREED THAT the announcement of my leaving the West Wing would be made the next morning. I also let Mark know that I wanted to be able to tell my comms and press teams in person before they saw it on the news; then the announcement would go out at 9:00 a.m. That morning, I had a specific timeline for the day: call Lindsay and tell her I had accepted her old job, tell Hogan what was going on, then sit down with my senior team—Hogan Gidley, Judd Deere, and Jessica Ditto. As I was telling them that I would be leaving, I got a text from a reporter saying that one of his colleagues had the story that I was leaving and asking me if it was true. On the one hand, I could not believe that was happening, but on the other, I was not surprised at all. I suspected Meadows and his team wanted to humiliate me and get it out there that I was being replaced by both Kayleigh McEnany and Alyssa Farah, not that I was leaving of my own accord. My team and I watched as the story broke, and I felt horrible that the other 90 percent of my staff was learning about it in that manner.

Mrs. Trump texted me immediately, "Do you see the news????," and I told her plainly that I believed the chief of staff's office was responsible, giving her the name of the person who had done the leaking, as told to me by three reporters.

Mrs. Trump obviously said something to the president, because about fifteen minutes later a red-faced Mark Meadows came into my office. "I have been told you are telling people my team or I leaked the story," he said, his voice rising. "Let me assure you, if I wanted to leak something on you, it would have been done long ago." OK, badass.

"Look, Mark, I believe your people leaked it," I replied. "I have never had a leak in all the time I've been here. You can look to the East Wing as proof of that."

"You need to be sure you are not telling people this was me," he said, "because that will be a problem." By "people" I assumed he meant the president. There was nothing Trump hated more than leakers—except, of course, when he himself was a leaker.

"Why would I leak that I was being replaced? We are going to have to agree to disagree on the series of events today, among other things, and just leave it at that," I said. "The difference between you and me, though, is that when I suspect something, I don't go storming into a person's office to complain. This is just how the place works." As he strode out of my office, he grabbed the doorknob and turned back toward me. His demeanor changed in an instant. In an eerily calm voice, he said, "I really am happy for you. Congratulations. This is very exciting." What did I tell you? The guy is strange.

During the next couple of weeks, his team wasted no time in removing any junior communications staffers who had worked for me, apparently convinced that they were "spies," and proceeded to treat most of my senior team with disdain. As mentioned before, promotions and raises that I had put into place were all rescinded. It was difficult to watch people who had been so loyal to the president and the administration be treated so poorly by the newest round of employees, who seemed loyal only to themselves. Even worse for me, those with any authority to speak up or say any-thing didn't. Of course, I hadn't always said anything when others had been treated the same way, either. After so much time in the Trump White House, you were just glad that you had survived the latest purge and maybe even got a weird thrill out of it. There was almost a sick pride in outsmarting those around you. When I say there was a reality TV show mentality to our administration, I'm not kidding. And I am ashamed to say that I lost sight of what I was there to do—serve the country—more than once. That was life with Donald Trump.

19

Dog Park Girl

I stopped waiting for the light at the end of
the tunnel and lit that bitch myself.

—UNKNOWN

I don't want to spend much time on this part. I just don't. But it is part of this journey, and it is a big part of why I ended up where I am today. It took me a while to understand why this part of my life is so relevant to this book—but in a weird way that I didn't understand for months, even years, I think it's crucial to understanding life in the Trump White House.

During my time at the White House, I fell in love with a co-worker, and we dated for almost two years. Actually it was the Music Man, referred to in an earlier chapter. He worked in the advance office, and we first got to know each other while working on trip coordination before the president was ever elected. Our relationship started slowly, as I have a hard time trusting people. But he was always there, always patient, or so I thought. He came to be my safe place and one of the only people I trusted with every aspect of my life. I tend to compartmentalize—it was impossible not to, considering my place of employment—but he knew stories about my family, my past mistakes, my insecurities, and my hopes for the future. Maybe it was the emotionally charged atmosphere of the place, I don't know. As our careers progressed (well, mostly mine, with me pulling him along), so did our relationship. We'd moved in together and become the owners of a wonderful dog. We were always together—always—and I thought that would never change.

But at some point our relationship began to deteriorate. I had seen signs of his temper in the beginning and as per my usual pattern had brushed it off as my fault or my being dramatic. By the time I got sick and was isolating in my own apartment, we were arguing a lot. When I returned to our shared home after my quarantine, he was acting strangely. He seemed to be angry at me for quarantining alone for three weeks. By that stage, I could do nothing right by him. I would wake up every morning shaking with anxiety about his mood that day. I rose one morning and walked into the kitchen for coffee. He was on the couch and asked, "Why are you even here?"

As I said, I had seen his temper before, and it had scared me because of my own childhood experiences—stories and fears I had confided in him, by the way.

Over the next two months, he seemed completely erratic and angry and spent a lot of time with a dog park group that we had met through a neighbor. The same neighbor also offered him cocaine on a regular basis. There was an attractive woman in the group, and the couple of times I insisted on joining him, I got a weird vibe between the two. I am not a jealous person, but wow . . . women's intuition is a real thing. I mentioned my feelings over and over and said that they seemed oddly close for having met in a random dog park group. He called me a psycho. I tried to understand and talk to him about why it seemed he was being so abusive, and he turned to stone. I even sat his boss down one day and asked if he would be sure to "keep him busy with projects." Without saying it outright, I wanted his boss to know that I feared he might be abusing drugs, which in my confused mind was the only way to explain his rages. You have to remember, after covid arrived, many people in the White House were working from home and therefore had a lot of time on their hands. Something was going on and my mind was racing trying to figure it out. In the meantime my spidey senses about the dog park girl stayed strong, though he continually called me names or just stormed out when I raised my suspicions.

At one point during that period, I woke up in the middle of the night and he wasn't in bed. I walked down the hall to a scene that I don't care to fully lay out in this book, but it involved him on our couch and on his phone FaceTiming with someone. His pants also seemed to be missing. I freaked out, yelled, "What are you doing?," and he shut the phone off immediately. I ran into the bathroom and locked myself inside. Not again. Not again, and not with this guy, whom I had fallen in love with, shared a home and dog with, and had let into my life unlike any other man before him.

He knocked on the bathroom door a few times and when I wouldn't come out called me a psycho again, told me he was just "looking at fitness models," that I was "being crazy," and went to bed. That night I slept on the nasty couch he'd just been "scrolling through pictures" on.

THE DAY WE BROKE up is a blur, yet it's crystal clear. I had ordered a ton of seafood from Joe's Seafood, Prime Steak & Stone Crab, a well-known DC establishment near the White House that we both loved. Unlike in my previous relationships, I was trying hard to keep us together and not run away despite how I was being treated. I was trying to forgive his behavior and move on because that is what strong people do. But sometimes it is not about being strong. A lesson I'd never quite learned from my childhood is that in the face of abuse, you should not feel compelled to stick around and try to be "strong" or just keep the other person happy. Holding one of the most powerful jobs in the nation and being viewed as a "badass" by my family, my friends, and members of my team, I was ashamed to admit to some of the ways I allowed myself to be treated at home. So I never did.

We were waiting for the food to arrive when he got a text. In a matter of minutes, he put on his jacket and grabbed a cigar and our dog's favorite ball, saying he was going to meet "the group." I asked if I could come along, but, as with so many times over the past two months, he instantly got angry. Standing at the top of the stairs and visibly pissed at my request, he pelted the ball over the railing at me, saying "Fine, whatever."

Needless to say, I stayed home. Again, in my mind I was just trying to fix whatever was wrong. After he left, I called my little sister and sobbed on the phone so hard that I couldn't get words out. She just kept repeating "Sister, sister, what's wrong?" I finally let out what I had kept secret for the past months, what I felt to be the cruelty of the man I loved. He was gone for a long time, and when he returned, he claimed that the police had run the group out of the park due to covid, so they had been to "someone's house." The seafood had long since arrived, and I continued to try and remain calm. He went to the bathroom, and his phone was open. What I saw stays private because it doesn't really matter anymore, but I called him out for that and many other things. I had dealt with his temper and behaviors for a very long time at that point, had been quarantined for three weeks and had then lived in a special kind of hell for two months—all in the name of "love."

The months of what I felt were abuse and cruelty had taken their toll, and I was finally done. When I left the house, it was a chaotic scene in every sense of the word, and I will leave it at that. I panicked, grabbed my purse, and ran out of the building. Finally, outside in the cold with no coat, I sat on a curb crying as I waited for an Uber. I texted my sister, "It's done. I left."

Once we were apart, I thought his cruelty would be over. I have never been more wrong in my life.

EVEN IF IT HAS happened to you, I'm not sure anyone is prepared for the pain of a volatile breakup. That night, my assistant, Annie, who had become a close friend and confidante and knew me better than most anyone, came over. She sat on my bed with me until I cried myself to sleep. That was one of the kindest and most comforting things a person has ever done for me.

The next couple of days I cried, then slept, cried, then slept. Friends stopped by to sit with me, I watched reality TV, rinse, repeat. A couple days after I left, my now ex-boyfriend sent me a text saying that if I wanted our dog I could have him, so that was at least a bright spot in an otherwise miserable time. Our dog, after all, had been my birthday present.

A week after the breakup, I started asking about the dog and received only silence. I heard nothing from him for four or five days, and then he finally sent me an email to say I couldn't have him. His reasoning was that *our* dog, who had lived with *both* of us since we had brought him home, couldn't leave his condo because it "was all the dog knew." He reminded me that his name was on the records and paperwork that we had (because I had always been holding the dog when those things were being filled out). He said he would "pay me back the half that I'd spent on him" so that we could both "move on." I'm sure you'll be shocked to know that he never paid me back—not that I cared about the money. I'm sure only dog lovers and pet owners will understand this, but I was devastated and destroyed all over again. In a matter of weeks I had lost another thing that I loved.

I waited a few days, then proposed that we split time with our dog. My ex said no to that, too. He had already begun a whisper campaign at the White House about what a "psycho" I was and that I was "demanding" to keep the dog. I even consulted a lawyer about ways to gain custody but eventually decided that I didn't have the fight in me. I was also reminded that if we did share the dog, I would be connected to an ex-boyfriend for years to come, and that certainly didn't seem feasible, either. I will say this, though—because (in my mind) it is important for later and is something I've never said out loud: the entire time that was all taking place, I thought we'd end up back together. Again, this is how abuse victims think, and there is no "off" switch for love.

PART OF ME REALIZED, of course, that my struggles were very minor considering that during that period hundreds of thousands of people were dying from covid-19. But I felt helpless about how to address the situation, and so, it appeared, did the Trump White House. As we all sadly know now, our isolation during the pandemic made some things in our own lives seem bigger or harder than maybe they really were. I think that was the case with my dog.

As Meadows tightened control over the White House operations, he appeared to be keeping more and more voices away from the president. Reasonable people who saw what was happening, gave Trump good advice, and sometimes told him what he didn't want to hear—Kellyanne Conway, for example—seemed like they were being sidelined. People who were suspected of being loyal to Mulvaney or who, I believe, Meadows deemed a threat were suddenly not invited to meetings. Department heads were told that they could not hire anyone without the chief of staff's approval. In fact, the Chief of Staff's Office started to dictate who could attend what meetings and events—no matter the level of importance. Meadows would bring in his favorite "Trump whisperers" to either calm the president down or give him pick-me-ups. They included familiar staples such as Lou Dobbs, Sean Hannity, and congressmen Jim Jordan and Matt Gaetz. We all knew that whenever Trump needed someone to defend him on TV on anything, Gaetz was our boy. He would do anything for Trump and a TV hit—though not necessarily in that order. When the president needed someone to tell him how awesome he was, the staff would get Gaetz on the line and he'd sing for his supper.

And, of course, the upcoming election influenced every decision Trump made about the pandemic. He had an almost mystical belief that his "base" would miraculously propel him to victory just like it had in 2016. So he listened to Brad Parscale, Jared, or any of the crew I just mentioned if they seemed to know what the base believed. The president also relied on pollsters such as Rasmussen Reports that always seemed to have the ego boost he needed—polling numbers that differed from everyone else's and always showed him in the lead, tied with Joe Biden, or moving in the right direction. Not a day went by when I worked in the West Wing that the president didn't ask Dan Scavino, "What does Rasmussen say today?" Trump seemed increasingly prone to delusion and conspiracy, and it looked to me that Mark Meadows was milking that for all it was worth. Why? Probably because that was how he stayed in power. On covid, Meadows and the whisperers nursed Trump's worst instincts. One of them told him that he could not wear a mask in public because it would show weakness and piss off the base, and everyone else blindly agreed. Because they probably knew what he wanted to hear, they encouraged him to be tough, which meant not showing empathy for the millions of people who were sick or afraid.

It is my belief that if Trump had behaved differently, if he had worked to soothe people's fears and maybe even once shown his own vulnerability, he might have changed people's impressions of him. At the very least he might have encouraged more people to take covid seriously, wear a mask, and thus saved lives.

I was reminded of what had happened in the fall of 2019 when Trump had made a seemingly impromptu visit to Walter Reed National Military Medical Center. I was informed, as were a small handful of others, including Vice President Pence, of the reason for the visit. Pence was told he had to stick around town "just in case." What I was not allowed to tell anyone at the time was that the president was having a very common procedure that all men and women over the age of fifty should have. In such a procedure, a patient is sometimes put under. In Trump's case that could mean signing a letter under the Twenty-fifth Amendment to put Pence temporarily in charge, but the president was ultimately not put under, I believe simply so he wouldn't have to be perceived as giving up power. I think in his eyes that would have shown weakness. I also suspect that he realized that the late-night shows would have a field day if they knew about the procedure—and I agreed with him on that. I would not want a trip to the gynecologist to become the butt of a joke on national television, either. Pardon the pun.

I thought the American people had a right to know about the health of the president, and I still do. But I didn't push the matter too hard. I think the president was embarrassed by the procedure, even though President George W. Bush had had the same one done when he was in office and had been very transparent about it. I had been so primed by the first lady to never, ever discuss private health matters of any kind that once again I made excuses in my head. But other presidents have been very public about their own bouts with illness or regular checkups and encouraged Americans to do the same. Trump could have used his own trip to the hospital to demystify the need for such a procedure and encourage others to have it done. In doing so, perhaps he could have saved lives. But as with covid, he was too wrapped up in his own ego and his own delusions about his invincibility.

A FUNNY—NOT FUNNY—STORY THAT happened in the middle of all this turned out to be used against me months later, but the story itself is still one of my favorites from my time at the White House.

Early in the breakup, my friend Rickie had stopped by to say hello and bring me a bottle of wine. She had connected me with two attorneys to try to get the dog back. After she left, I went to see my best friend, Art, and his wife. As had become par for the course with me, I was crying to them, asking for advice, and just generally moping. I was supposed to stay over that night, but as during every night since the breakup, my mind was racing and I couldn't sleep. There was so much happening in my life, and in the world, and I felt helpless. I got up and called an Uber at about 1:00 a.m.

On the way to my apartment, I sent Rickie an email with a list of points to make in case she spoke to the second lawyer before I did. I told her that I had screenshots showing proof of ownership of the dog on my phone and that I would be in the White House Situation Room for a good part of the following day, so I would leave my phone on my desk. I also gave her my phone's security code. When Rickie had been visiting earlier, she had suggested that I take an Ambien to help me sleep, so when I got home I decided to do just that.

The next thing I knew, I awoke in my bed with two men standing over me, asking if I was okay. Now, to get a clear picture of
this whole scenario, you need to know that I sleep with my TV
and a fan on, and that night, I had also turned on a stupid machine
that makes stars move across your ceiling. (That had been an "I'm
depressed" Amazon purchase the week before, so please try to go
easy on the judgment.) It's impossible to explain how disorienting
it was to suddenly wake up with all that going on around me and
see two strangers in my room.

After a minute, I was able to say that yes, I was okay, and then
ask what was going on. My first thought was that something terrible had happened concerning national security and I was about to
be taken somewhere secure. One of the men asked if I would go
out to the kitchen to talk. I threw on a robe, then walked into my
dining room. I could see a man in a suit in my kitchen, and there
were people in my hallway and another man in the dining room.
I didn't have my contacts in but it seemed from what the last man
was wearing that he was part of the Uniformed Division of the
Secret Service.

We all just stared at one another for a minute, then the guy in the
suit asked if I knew who he was. I walked up to him and got very
close (again, no contacts) and realized that it was Mrs. Trump's
lead agent. "Oh, hi! What is going on?" I asked the agent, and
he moved me to the dining room. He asked why I hadn't known
who he was at first, and I explained that I didn't have my contacts
in. He then asked why I hadn't answered the door, because they
had been pounding on it for some time. I reminded him that he
had just been in my bedroom and could see the circus of sounds
and lights that I had going on in there.

The other guy then asked if I had a dog in the apartment, which
I thought was weird, but so was the whole situation, so I said no.
The agent then pressed me as to if I had taken any drugs that

night. I said yes, I had taken an Ambien, as I hadn't been sleeping. Then came the kicker: he let me know that "someone" had been worried about me and had called in a wellness check. I was stunned. My mind was still foggy from the deep sleep I had been in, but I immediately started questioning myself—had I been acting so dramatically that someone had thought I wanted to end my life? I asked Todd who had made the call, and he wouldn't say, but then it hit me: my dear friend Rickie. She had clearly gotten my email with all the information about the dog and misinterpreted something.

Rickie was exceptionally caring but at times could take things a little far. When I asked if it had been her, they confirmed it. I went on to tell them the story of the email I had sent her and why, and that I was going through a nasty breakup, so this was all a very big misunderstanding. I was mortified, but I knew the agent pretty well, so that gave me a certain level of comfort. He assured me that only Tony Ornato, the deputy chief of staff for operations and his boss, knew about it, so I had nothing to worry about in terms of its getting out. They left, and I went back to bed actually chuckling a little bit. It was a weird story, and I probably looked crazy in it—but at least no one else would have to know. Right?

A few hours later, my phone rang. Mrs. Trump was calling to ask if I was all right.

I asked her how she could possibly know what had happened.

She said that Mark Meadows had called to inform her that he'd had to "send agents to my house to check on me." Well, just fucking great, this guy I hated and who I was sure felt the same about me now thought I was suicidal. I had no idea if the full story had been explained to him, and I was pissed off that he knew in the first place. The very fact that he'd called Mrs. Trump and played the sympathetic role was complete bullshit, and I wondered who else in the West Wing would know by the end of the day.

I told her the real story of what had happened, and knowing Rickie the way we both did, we laughed about it. She assured me that Mark had told her that no one else knew and it would stay that way.

Now that Meadows was involved, though, it was no longer a weird, funny story to me. It was humiliating and something that could really damage me professionally if it got out. Not only could it annul my security clearance, it would be just the kind of information that would be great fun in the gossip circuits all over DC. My next call was to Rickie to let her know I was okay and find out exactly what had happened. The long and the short of it was that Rickie had woken up in the middle of the night and seen my email. She had remembered that she had brought me wine and told me to try an Ambien to sleep. She'd had a cousin who had overdosed on Ambien a few years prior, so she'd gotten worried and called to check on me.

When I hadn't answered my phone, she'd panicked and called the access control agent at the White House, whom she knew and trusted. She'd told him she was worried and asked if he would go check on me—not realizing that he would have to alert his superiors and set an entire process into motion. She felt truly horrible and offered to call FLOTUS, Meadows, Tony Ornato, and the agents who had checked on me to let them know it was "her fault" and that she had overreacted.

I told Rickie kindly that maybe we'd had enough phone calls from her to the White House about me. I know she was just trying to help, and I actually loved her even more for it. I laugh about it with her today, but I think it is still too soon for her.

I sent both the agent and Tony an email apologizing for the drama and thanking them for their concern and discretion. I mentioned that it was nice to know that we were all looking out for one another. Tony stopped by my office later so we could laugh it out, and he assured me that there was no paper record of the incident and no one knew but Meadows and it would stay that way. But they didn't know Mark Meadows the way I did and when I told him FLOTUS had been told he seemed surprised.

Later that evening, Mrs. Trump emailed me to say that POTUS was worried and had asked about me, because surprise, surprise, Meadows had felt the need to inform him of the incident. She assured me that he knew it was all a misunderstanding and not to worry, writing that they "both loved me." That made me feel better. The president and first lady had my back, I thought, after my experience with terrible abuse, drama, and trauma. I didn't realize then that I was just continuing a pattern.

20

Boys Will Be Boys

The bad guys don't wear signs. And all of us are only human.

—FROM *BLOOD ON THE TRACKS* BY BARBARA NICKLESS

I should probably say at this point that the president and first lady had known for a long time about my boyfriend. I'd helped him get plum assignments on advance trips to places such as Afghanistan and Iraq, and, most important to him, I'd put him in front of the president as often as possible.

I think a lot of women will get this when I say that I was useful to him until he'd gotten all he could from me. Stephen Miller once put it that way, apparently telling my boyfriend, "You got in when the stock was low"—meaning that he had gotten to know me when I was still a more junior staffer and then ridden along with me as I'd gone up in the ranks.

Now, nobody in the White House loved gossip more than Donald Trump. After the president found out that we were dating, he constantly asked me how things were going and told me what a good-looking couple we were. My boyfriend even told me that on an Air Force One flight, as they had been listening to music, Trump had asked him if I was good in bed.

Shortly after our breakup, Trump, my ex, and I found ourselves together on Marine One. Nothing awkward about that! But to make it a bit more awkward, the president leaned over at one point and whispered to me, "Hey, how was it seeing your ex tonight?"

"It was okay," I replied, trying to be professional and praying that that line of questioning would end.

"Oh, this really hit him hard," Trump said. "He's really upset."

"That's bullshit," I replied, my temper flaring. You know, maybe I didn't need to get into it with the president of the United States, but if he had asked how I was in bed, I could at least give him the down low on what my relationship with his new buddy had really been like. "He was terrible to me," I said. "He cheated on me and lied to me."

"What?" Trump said, passing no judgment on any of it. "He told me he hoped you two would get back together."

"That is bullshit," I said again. "He's just saying that to look good to you." I tried to keep our voices low, since my ex was only a few seats behind us.

I also told the president that at the end I felt my ex had been abusive toward me, information I had shared with Mrs. Trump as well. I know that it was my word against his but I wasn't about to sit there and hear how hard things were on him.

That also seemed to have no effect on the president. It occurred to me that perhaps he wouldn't be so eager to tell me what a bastard my ex had been for cheating and treating me so badly; after all, the first lady was sitting there, part of the conversation, and no doubt he was hoping not to remind her of his own sketchy past with women (as if she would forget).

THE PRESIDENT HAD HIS own vices on the job, if I could put it that way. When I was press secretary, I began to notice that he was taking an unusual interest in a young, highly attractive press wrangler on my team. Yes, the same job I'd once had, which now seemed a lifetime ago.

Let me be clear before I start: I do not know if and am not alleging that anything happened between the president and that woman. It would have been pretty hard for him to pull that off even if he'd wanted to. What I do know is that he behaved inappropriately. And since the woman worked for me, I tried to protect her and keep his unusual interest in her under wraps.

As I mentioned earlier, the press wranglers are in charge of escorting reporters into and out of the Oval Office and events and generally trying to keep them in line. Trump noticed some of those people, as he had me all those years ago. But that particular young woman he noticed way more than the others.

If the president didn't see her with the press corps, he would ask me where she was. He would ask me if she were coming with us on foreign trips. When she did come along on trips, he often asked me to bring her to his office cabin in the aircraft, which he'd rarely done with anyone else. Sometimes I would make an excuse, but on the occasions when I couldn't find a way out of it, I always accompanied her and stayed in the cabin the whole time. My instincts were on full alert. The whole thing never felt quite right.

When Mark Meadows brought in his new press team, they let some of the wranglers go. Trump liked one of them, a tough, loud woman who had been with him since the beginning of the administration, let's call her Suzanne. Kellyanne Conway often looked out for the junior press staffers, and she went to the president and told him, "You need to know they fired Suzanne. These are junior staffers who have always worked very hard for you, and this kind of treatment shouldn't be tolerated." She was that aggressive in the hope that Trump would care and direct them to keep Suzanne. He didn't. The only thing he wanted to make sure of was that Meadows hadn't fired the other woman, the one he seemed to find so attractive. "She still works here, right?" he asked.

On Air Force One, I was in constant fear that the press would start to notice how often that Trump requested she come up. Usually press wranglers stay in the back with the press. When I accompanied her to the office cabin, I never said much other than he had requested to see her. I didn't actually know her that well, nor was I sure that she even sensed that Trump's attention was unusual. I certainly didn't want to freak her out or have her start talking about it with the press.

"Put her on TV. Keep her happy, promote her" Trump would tell me. He even said that to her. "Do you want to be on TV? You'd be great on TV, a real star," which of course is the highest compliment in the Trump universe.

On one trip that my deputy went on in my absence, Trump asked how the woman was doing before instructing them to bring her to his office cabin. I got a call afterward relaying that the president had said, "Let's bring her up here and look at her ass." After that, I tried to keep her off trips. I didn't want to punish her for something that as far as I knew she had no part in. But I also knew I needed to protect her and, frankly, the president as well.

There was one other option I considered. A couple of times I came close to telling Mrs. Trump about the president's behavior. I thought that if she would say one little word to him about it, she could make it stop. But I could never bring myself to say anything. Maybe I just didn't want her to have to go through all the shit I was going through. Maybe she wouldn't believe me. Maybe she would believe me but blame me for telling her or think I was overreacting. Maybe I was too chickenshit. Maybe it wasn't my place or any of my business. I don't know. Mrs. Trump still may not know.

*   *   *

MY LITTLE SISTER WAS a great source of strength for me when I had bad days at the White House, as she had been since day one. Suffice it to say that our phone calls became daily after the breakup and as my issues with Mark Meadows continued to fester. One day she suggested that I hop into my car and drive to see her in her tiny Kansas town—"It's only a twenty-hour drive," she said. The way she talked about it, the whole thing would be like making a quick drive to the grocery store. The town has a population of 1,400 people and boasts four restaurants. I'd been there to visit her before, so I laughed her suggestion off at first. Then on another of my sleepless nights I got up and packed a bag for a "long weekend." The next morning, I loaded the bag and some snacks into my car and hit the road for Kansas. I took one other thing—a new ten-week-old French bulldog named Ben, who was well on his way to becoming my new best friend. The road trip was an adventure, to say the least—and a twenty-hour drive, by the way, is a very long drive (thanks, Sis). I hadn't driven a vehicle in a very long time, and Ben was far from being potty trained. I learned to navigate bathroom breaks at rest stops by taking him inside with me hidden in a large bag. It took us three full days to get there, but we made it and I surprised myself by just how much I enjoyed the open road and blasting music on the car's radio.

By that point with covid, the East Wing was telecommuting 100 percent—as I mentioned, Mrs. Trump had basically been telecommuting almost from the beginning anyway—so I was able to take calls and answer emails. Arriving in that small town was like taking a breath of fresh air that I hadn't been exposed to in almost six years—literally and figuratively. The people are kind, unassuming, and willing to help you, and there is no judgment. Life there is so simple that it was actually hard for me to understand it at first. By and large, everyone gets up, goes to work or school, comes home, and makes supper (not dinner, I soon learned), has drinks on the porch with neighbors, watches some TV, and goes to bed.

Turns out that when I say simple, I meant simple for me, because I am describing the lives of probably 75 percent of American families; I had just not lived a "normal" life for so long that it felt foreign to me. Being with my family and far away from the swamp that is DC was just what the doctor ordered. I played games with my family and slept a lot, Ben was able to run around in a big yard, and I ate home-cooked meals every day. I began to understand why people remove themselves from situations to regain perspective and appreciate things they have been neglecting.

A long weekend turned into a week, and with each day that passed I dreaded going back to DC more and found a reason to stick around another day or two. I made plans to see other members of my family. I kept my whereabouts a secret from most everyone. I didn't need the rumor mill grinding any more than it had been, and I wanted to hold on to the peace I had found through privacy.

I HAD BEEN IN Kansas for a couple of weeks when I first heard that Mark Meadows was accusing me of leaking "the bunker story" to the news media. About a week before I'd left for Kansas, at the end of May 2020, Black Lives Matter and other protests had become a daily occurrence in the nation's capital. On Friday, May 29, hundreds of protestors had marched close to the White House, throwing objects and shouting menacing things at President Trump. The situation had become so tense and uncertain for a moment—especially after some protestors had breached the White House's outer gates—that the Secret Service had decided to move the first family to a protected bunker as a precautionary measure.

I was not in town that weekend, but, as was standard proto-col, Tony Ornato, the deputy chief of staff for operations, called me after the first family had been moved to let me know that Mrs. Trump and Barron were safe. He said it was just a precaution and that they would all be leaving to go back to the residence soon. The next day, I started getting calls and texts from reporters asking me if it was true that the family had gone to the bunker. Cell service was poor in the area I was staying in, and I was hon-estly relieved to have an excuse not to engage with the press on that one. I did send Ornato a text, though, to give him a heads-up that word had gotten out somehow and that obviously I would not be responding to the queries. When it came to matters such as national security and the first family's safety, there was never a chance in hell that I would give the press the slightest bit of in-formation.

When the stories finally did come out—the first one appearing in the *New York Times* that Sunday—the president was apparently furious. He felt that the image of his going to the bunker made him look weak or afraid—and he hated the very thought of some left-wing protestors taking satisfaction over making him run scared, no matter that it was the decision of the Secret Service. To Trump, as I've stated many times, there is nothing worse than being made to look "weak," and he denied that he had gone there for his safety, saying that he had gone for an "inspection," which was just not true. He then went on a tear to hunt down whoever had leaked the story, apparently tasking Meadows with the job. If you ask me, that was like putting a pyromaniac in charge of fire safety, given that Meadows and his team seemed to be leakers themselves.

I suddenly started getting terse emails that Meadows needed to meet with me "immediately and in person." As would become a trend—and my saving grace in many situations—toward the end of my time at the White House, it was reporters who were kind enough to tip me off to what was being peddled about me inside the White House and outside to the press. I got a couple of calls letting me know that Meadows and his team had apparently been telling people that I had leaked the bunker story to the media and was going to be fired.

WHENEVER WHITE HOUSE OFFICIALS undertook a hunt for the leaker of one story or another, the leaker was rarely found. But the search was often used as a pretext to get rid of people they didn't like. The most notorious example was the hunt for the anonymous White House official who had written a *New York Times* op-ed and book stating that Trump was morally unfit and a danger to the country. Peter Navarro had put himself in charge of that investigation, and he and a few others had named an NSC official named Victoria Coates as the prime suspect, based on a bunch of bullshit that they claimed was evidence. She was innocent, but they didn't care. It seemed they wanted Coates out of the White House for various reasons, and her being the anonymous writer was the perfect excuse. Victoria went through weeks of hell, which included people leaking about her to the press and effectively turning her into a nonperson within the walls of the White House. Eventually, the Department of Homeland Security's Miles Taylor revealed himself to be the anonymous writer. I don't know why Meadows thought I had leaked the bunker story. I'm not sure if he thought it was true or was just using it as a way to get rid of me, but as with all such situations in our administration the label of "leaker" stuck on me in the eyes of at least a few folks.

A conservative columnist, Miranda Devine, wrote in the *New York Post* that according to her sources, only three people could possibly have known that the first family had been moved: Dan Scavino, Nick Luna, the president's body man, and me. That was preposterously untrue. For example, members of the Secret Service knew, because they were the ones who had moved them. Tony Ornato knew, since he had called me to tell me the news. Anyone who had been sitting around the president in the Oval Office knew, since they had all seen him walked out of there. Any staff members who had seen the first lady or her son on the move also knew. That's easily a dozen people or more right there. But the suggestion that I was a prime suspect was all Meadows needed and Ms. Devine gave it to him. I know this might sound paranoid to those who didn't see firsthand how that group operated, and so be it, but I suspected an effort to find a way to get rid of me once and for all, and I wasn't having it.

The back-and-forth via email between Meadows and me was once again tense, to put it mildly. He kept requesting to see me. I would say that I'd be happy to talk on the phone because I was telecommuting from out of town. Then he would say no and that "it must be in person." I would ask him what he wanted to talk about, and he would just ignore me. All the while I was keeping Mrs. Trump looped into what was going on and what I was hearing. She offered to call him for me at one point, and I declined, though I did ask her to make sure that the president knew what was going on. I had been out of DC for three weeks when I once again got an email from an Axios reporter telling me he planned to write that Meadows was telling people in the White House that I had leaked the bunker story, that Meadows had told the president "what he knew" and recommended that I be fired. He went on to say that he was going to report that Mrs. Trump had "saved me" from the humiliation of being fired.

I was furious and forwarded the email to Meadows, writing "I can only guess this is why you have been asking to meet with me." Keep in mind that it was the second time I'd had to forward an email to Meadows requesting he go on the record to defend me. I don't believe that was a coincidence. I wrote, "I have been with the Trump family for almost six years and have never betrayed their trust, and I certainly don't plan to. I'd like to remind you that I gave Ornato a heads up when reporters started asking me if the bunker story was true, and I have screen shots to the effect should you want to see them. I expect that you will go on the record with Axios to clear this up immediately."

My email was met with silence. I found out later that members of his staff had vehemently denied having started the rumors and had offered to go on the record to say it wasn't true, so that was something. But after covid, my quarantine, my breakup, my very public job change, and the bullshit with the new West Wing chief, I had zero fucks left to give over Mark Meadows and his mean-spirited (and amateurish) maneuverings.

But I'd be lying not to admit that it was all taking a huge toll on me. I was physically sick most days from the anxiety of dealing with what I perceived as harassment by Meadows and his minions. So I sent Mrs. Trump an email (once again) offering to resign. I wrote that if she wanted me to stay, I refused to work with Meadows in any capacity.

She called me immediately. She said, "You do not resign. You must be strong and not let him bother you. The president and I know you do not leak, we will just do our own things like we do in the East Wing. You do not have to talk to him. Emma can take meetings."

I appreciated her support and all she was doing to fight for me, but I also felt tired and disappointed. Even though I'd opened the door to staying in the administration, I really wished I'd quit. Here again was the sick pride I was holding on to, feeling that my worth derived in large part from having been a loyal "survivor" over three and a half tumultuous years at the White House and countless comings and goings of personnel. I had outlasted three chiefs of staff and wanted to outlast the rest. Part of me just couldn't let Meadows win.

You can feel sorry for yourself or angry for only so long. Now that I knew I could telecommute and travel back and forth to DC when needed, my sister and I came up with a bonkers plan. The house across the street from her was for rent, and I could think of nothing better than to create a country home of sorts as an escape. I would be far away from the swamp while still doing my job and being close to family. My sister and I drove to DC with Ben, packed up my apartment, and headed back to Kansas—all in three days. I told no one what I was doing except my closest circle of friends, because at that point I was paranoid and looking over my shoulder all the time. The newest narrative in the press was that I was "checked out," and the new living arrangement certainly wouldn't have helped that storyline. Looking back though, I was checked out. That was true. A person can deal with the pile-on of hatred for only so long. I should have checked my pride and giant ego at the door and left when Meadows started.

I wish I had stuck to my guns and resigned when I'd tried to, especially considering all that happened next.

21

Snakes and Home Depot

In the end you have only yourself.

—SUNNY RAWSTAR

As I grew accustomed to my new setup, I decided to start making small improvements in and renovations to my little rental house. I thought manual labor would be, well, a sort of therapy. The only issue was that I barely knew how to hang a picture on the wall, let alone carry out all the ideas I had. I went to Home Depot almost every day for two months. My sister and I yanked all of the carpet out of the house by ourselves—and we're talking truly gross, heavy, nasty carpet. I also encountered many snakes and found myself on the top of chairs, deck railings, and anything else off the ground while screaming for my sister to "just get it away!" I'm sorry, snake lovers, but anything that can fold itself in five places at once while it slithers around is just not for me.

I sanded floors with a hand sander, including one memorable day when I was missing and hating my ex so much that I worked nonstop for seven hours on my hands and knees, sanding the living room while listening to music, sweating, cursing, and crying. I also tried to replace light fixtures until I got shocked off of a ladder; I didn't know that you need to turn the electricity off first. I painted every wall in the house and hand sanded all the door and window moldings, along with the kitchen cabinets—all of which had easily seven layers of paint on them from previous renters. I went back to my sister's each night with swollen knees and cuts and bruises everywhere. The house became a labor of love—or hate, I don't know which.

One day, furiously sanding my floors as if I were taking my anger out on the wood, I suddenly stopped. I'd heard before about people having an epiphany, a moment when everything suddenly becomes clear. That was what happened to me. Suddenly I got it. It all made sense. All my life I'd dealt with abusive men, people who had lied to me, berated me, made me feel like shit. They'd kept telling me that I was crazy, I was psycho, I was emotional, and I couldn't believe my own eyes and ears. And I'd believed them, or I'd wanted to. I'd kept forgiving them, kept trying to get them not to abuse me physically or verbally. When that didn't work I would convince myself that I was a bad person and that the way they were treating me was because of something I had done.

The Trump White House was just a fancy new setting for the same old crap I'd dealt with my whole life: my ex, Mark Meadows, even the president—all men who I felt had lied to me, lied about me, called me names, and/or made me feel worthless. With my ex and with President Trump, I'd done everything I could just to make the pain stop, to get them to like me again, accept me, be happy. But Mark Meadows—the hell with him. He'd come along too late in the game for me to care about what he thought of me. From that moment, I was done with the West Wing of the Trump administration, even though it wasn't done with me.

THERE WAS STILL MY job for the first lady. I still liked her, respected her, and wanted to help her. She had almost always been good to me so I did my best to put a mental bubble at least around my work for her. In the final months of 2020, that was pretty easy. Even though there was a heated reelection campaign going on, with a global pandemic raging and her husband doing poorly in the polls, the East Wing was like it always had been—chill and often inert. But I do remember the tennis pavilion.

Mrs. Trump worked for months with her designer to remodel the tennis pavilion at the White House. Critics would comment that it really wasn't the most urgent priority at the time—fair enough—but she did a damn good job. As usual, she went through all of the photos afterward, including of the small opening ceremony. For whatever reason, a photo of just her with the ribbon in front of the new pavilion had not been taken. In a situation like that, most people would throw up their hands. "I wish that picture had been taken, but oh, well!" Not our girl. Mrs. Trump requested that a new ribbon be created, and a few weeks later she went back to the pavilion with her photographer, a couple of staffers, and the chief usher for a two-hour photo shoot—you read that right—and you can be damn sure it included multiple photos of her cutting the new ribbon, sitting inside the pavilion and posing as if she were looking through the coffee table books, and on and on.

In other words, we were getting back to normal. I'd often seen Mrs. Trump do that. On holidays or before a major event at the White House such as a state dinner, she would summon her photographer and take photo after photo, posing and walking around, inspecting the decor or straightening dishes. Then she would survey them all, like Howard Hughes with his fingernails, except in this case photos on top of photos, albums of photos. Maybe they were her comfort food—and Lord knows we all needed it in a nuthouse like that one.

I don't want to paint a picture that the first lady didn't care about policies or her husband's success. It wasn't quite that simple.

She brought on board my friend from Mulvaney's office, Emma Doyle, to help with policy ideas. She did still care about Be Best and children's issues. She would ask on occasion how the polls were looking for her husband and what states it might be helpful to visit. But in that final year, she mostly canceled whatever plans we started to make. At one point we tried to get her interested in Native American issues and taking a trip to Arizona—another state crucial to the president's reelection. She said her usual "Let me think about it," which meant no. Like her husband, she didn't like to go places that were far away.

As the walls were closing in on her husband's presidency and with the pandemic upending everyone's lives and freeing her of most of her official obligations, Mrs. Trump seemed to retreat further and further into her own interests and priorities, which included making photo albums of every holiday, major event, and special moment. Maybe she saw the writing on the wall and wanted to preserve the perfect image of herself for posterity, maybe she was just bored, maybe she just didn't care anymore and was going back to her model origins, obsessing over her image. I'm not sure what she was thinking, and she did not share it. It made me feel sad and relieved and annoyed all at once. But I was also so wrapped up in my own personal and professional dramas that I think I was selfishly grateful for the reprieve.

Still, the Republican National Convention was coming up, and I wanted Mrs. Trump's speech to kick ass, so I put a lot into making sure that it had substance (and was 100 percent original). The location of her speech was a whole other story. The campaign kept giving us mock-ups of places where she could deliver her speech, one of the options being the beautiful Andrew W. Mellon Auditorium on Constitution Avenue. Ever the model, she wanted the venue to be outdoors if possible, because of the lighting. I reminded her that the speech would be at night, but she still insisted on being outdoors. When she finally selected the newly renovated Rose Garden, I vehemently disagreed.

THE ROSE GARDEN RENOVATION had led to another media controversy and another big headache for our office. In August 2020, the first lady unveiled the redesigned Rose Garden. She had used the plans of the original designer, Bunny Mellon, to make the changes, but that didn't matter. The internet pounced. The new garden was denounced as "sad," "pale," even racist. One columnist in *USA Today* said it was "a metaphor for the Trump administration—no diversity, no color, a parade ground framed by white columns only." If that's a metaphor, it's a tortured one. The Reuters fact-checkers chased down another absurd claim: "Tens of thousands of social media users have published posts claiming that in her redesign of the White House's Rose Garden, First Lady Melania Trump has removed roses from every First Lady since 1913, including Jaqueline [*sic*] Kennedy's crab apple trees. This claim is false."

Never mind that the trees Mrs. Trump had removed (which were not Jackie Kennedy's trees) were killing everything around them and that changes had to be made to make the garden more accessible to visitors with disabilities. Never mind that she had introduced the first work of art by an Asian American into the White House art collection by placing Isamu Noguchi's sculpture *Floor Frame* into the Rose Garden. The press just wanted to scream about racist columns. It was the media at their laziest and most unfair.

ASIDE FROM POTENTIALLY REVITALIZING all the controversies about the location itself, I thought that if Mrs. Trump spoke in the

Rose Garden, her speech would be overshadowed by the fact that we were using the People's House for an overtly political event. Usually she was highly sensitive to such things. Plus, I knew we would destroy the brand-new grass that had just been put in. She dug in (have you noticed a pattern here?) and I lost that battle, but the speech itself ended up being all I had hoped it would be.

Mrs. Trump even practiced it a few times, which was not something she normally did. The evening of her speech, I watched her give it from the front row, and the accolades that came in were well earned. She did great and had even allowed some more personal stories to be integrated into her speech. After it was over, I met her in the Cross Hall to let her know the reaction it was getting online. The president was with her, and I was beaming when I said, "Ma'am, you killed it. You are getting rave reviews!" Ever her own worst critic, she pointed out a word that she had stumbled over, and I told her not to even think about it. The president then gave me the first and only hug I would ever receive from him, telling me that the speech was wonderful. He then told Mrs. Trump about me, "One more like that, and I'll have to take her back." Mrs. Trump laughed and said, "She is never going back to the West Wing," and I replied that that was very true. That, as I recall it, would be the very last time I spoke to President Donald Trump.

THOUGH I WAS ONE of the first to be impacted, covid hit the West Wing hard. Junior and senior staffers, a member of the president's valet team, and even Meadows tested positive—as did the president and Mrs. Trump. On October 1, I learned that they had tested positive and immediately texted her, asking if she was okay. "I am ok. I should put something out on tweeter tomorrow morning?" Tweeter, by the way, was how she referred to Twitter. I replied,

"Yes, I will draft you something," and we talked about canceling some events that had been planned in Colorado and Oklahoma. Dr. Conley put a statement out that night, so I suggested that she tweet, too, as I was being inundated with questions. The next morning, I sent a text to check on her and also to let her know that I was being asked about their son, then finally giving her a heads-up that many reporters were hearing that POTUS was having symptoms in case he wanted to address it. I wrapped up by saying "Please keep me posted on how you're doing and I'm thinking about you both, here if you need me. Xoxoxo." She told me that I could tell people that their son had tested negative and was fine.

The next few days she was quiet, which I took to mean she wasn't feeling well. When she did respond, it was always "I have mild symptoms, body ache, temperature on and off. Who is asking?" She always liked to know who was asking what. Over the next few days, we didn't communicate much—she was choosing to go a more natural route in terms of medication, so her recovery took longer and she understandably wasn't talkative. I was asked to do TV several times to talk about her recovery, and she always told me no. In terms of the president, who had been taken to the hospital, she related only that he was bored there.

In the midst of all that, one thing that struck me was Mrs. Trump's anger over Stephanie Winston Wolkoff, who had recently published a tell-all book about their relationship and was also releasing audiotapes of their conversations. Mrs. Trump sent me a text saying "Your thoughts? That I do an oped on the tapes she put out. To set a record straight. If not people will not know the truth." I responded with a pretty lengthy text that said, in part, "There's nothing wrong with having one ready. I think timing is very important though. You and POTUS have covid, and if the first thing you do is file a lawsuit with the Department of Justice and write an oped about this, it may seem self-serving. And she is not being looked at with pity by anyone." We went back and forth for a few days, her wanting to push back forcefully, me wanting her to lie low and focus on covid. On October 13, the Justice Department filed a complaint against Ms. Wolkoff, claiming she had breached a confidentiality agreement by publishing her book and asking for any profits from the book to be put into a government trust (the suit was later dropped).

To his credit, the White House counsel had cautioned Mrs. Trump against using the DOJ to do that, but once again, she had dug in and made her wishes known. A few days later, we posted a blog on the White House website that came from Mrs. Trump personally. It attacked the media's coverage of the book rather than her good work and the release of the tapes, among other things. Thankfully, she had agreed to send out an update on her health and recovery the day before, so our response to Stephanie didn't seem quite so insensitive as the virus continued to make its way across our country and around the world.

22

Election Night

Pour yourself a drink, put on some lipstick,
and pull yourself together.

—ELIZABETH TAYLOR

I had a strong feeling we were going to lose the 2020 election.
I say this as someone who had been one of the few who was
convinced that Trump would win in 2016. But this time was different. It wasn't our lackadaisical and confusing response to covid
that convinced me. It wasn't Joe Biden. Long before the pandemic
struck, I saw things going against us. I had a feeling that the election wasn't going to be about policy this time around. I just figured that people had become tired of all the controversies and
scandals and Trump's offensive tweets and statements. In short, I
thought Trump had exhausted everyone.

I didn't want him to lose. I was a conservative who believed in
many of the policies the administration put forward. I didn't want
my friends and colleagues to lose their jobs, and I was not on board
with the plans Joe Biden had for our country. On the other hand,
I didn't want Trump to win, either. It was as though I wished we
had another Republican candidate, which was of course, impos-

sible. In my mind at that time, so much damage had been done by the Trump administration—and I am not talking about policy issues. So much ugliness had been unleashed. It's not as easy as saying "His tweets were too mean." It's not that black and white, and to this day I find it hard to explain. He hadn't changed much in four years, and I was no longer sure that that was a good thing. He was still too often president of his base and not the country. That had been noble in the beginning, when he had truly been fighting for those who felt they had no voice, but after he was elected, I think his priorities should have expanded to include the entire country—even those who disagreed with him. He missed opportunities to reach out, cool things down, and empathize with people who were hurting or who didn't like him. Being strong is important, but so is being honest and humble. In my mind, our administration had become about one man and who was or wasn't loyal to him. I feel that we lost sight of our country, and amid all the noise that was covered 24/7 by the media, the achievements of the administration were being largely ignored. Because of all that and more, I didn't know if Donald Trump deserved reelection. So I was dreading election night because I wasn't sure I'd be happy with the outcome either way.

AS WE SETTLED INTO the witching hours of election night 2020, nearly 1:00 a.m., the White House watch party downstairs was in full swing. Trump World figures and assorted hangers-on mingled with conservative media personalities and others—a *Star Wars* cantina scene as produced by One America News Network. But the good mood based on the night's early victories in Florida and Ohio started to turn gloomy as Joe Biden gained in the vote count. The top-tier advisers to the president—Jared and campaign manager Bill Stepien, among others—went upstairs to the residence to try to get the president's ear.

But where was the first lady? At the moment, nobody knew.

I knew that if President Trump was preparing to make an election-night speech with so much of the vote count left to go, the first lady would want to weigh in.

But she'd gone dark. She wasn't texting me back. That was unlike her.

She hadn't appeared for the election-night party, but we hadn't expected her to. In fact, she didn't feel the party was appropriate and had made that clear when it was being planned. For one thing, the first family had only recently gotten over their own bouts with covid, and she didn't want to be responsible for another perceived "superspreader event" in the White House or the message it would send to Americans. She also felt it was inappropriate to mix the election with the People's House—never mind that the entire Republican National Convention had been at the White House, including her own speech. But I think the real reason was that the big, splashy election party was the brainchild of Ivanka and Jared, and I think Mrs. Trump was frankly tired of the Trump kids thinking the White House was their personal playground or a Hollywood set on which they could perform.

Though the East Wing was consulted about the plans, after all of the years of our office tangling with her husband's children, the first lady didn't really think that Jared was asking for her permission or even cared what she thought. "They were planning this already, and they hope I will say yes," she told me. She was right, of course. Planning the election-night party became yet another test of wills, patience, and endurance among our office, campaign staffers, Javanka, and the president's third wife.

The campaign wanted to build a monstrosity of a stage in the

East Room for his acceptance speech; it had a huge guest list and equally large plans for a setup that would have been impossible in the White House, which is not only a private home but also a functioning museum. No one seemed to care about preserving the historic building that is the White House or considered the security plan if riots broke out in Washington that evening. I myself was concerned that we'd end up with a four-hundred-person slumber party at the White House. I remember that after the party started, Jared walked up to me and pointedly said, "I think this is all fine—I don't know what the fuss was about." Okay, buddy, noted.

Eventually, we made it work. The two sides ended up compromising on lighting, stage size, food selection, and number of guests. We removed every piece of historic furniture from the State Floor, and we made sure that everything we brought into the residence was signed off on by White House curators and residence staff. Covid testing of all the guests was to be conducted off-site. After the guests tested negative, they would be given a yellow bracelet and put onto a bus to proceed to the White House.

THE EVENING BEGAN JOVIALLY enough. There was plenty of space between guests, and TVs had been set up throughout the State Floor, including in the State Dining Room, the Red, Green, and Blue rooms, the Cross Hall, and the East Room, where the stage had been built. Tables had been set up, and there were two open bars. The buffet consisted of all of the president's favorite foods, which meant, among other high-calorie offerings, cheeseburger sliders, french fries, and chicken tenders. Because the first lady had made clear that she wasn't interested in attending, we hadn't made any requests about the menu on her behalf. Suffice it to say, she would have been disgusted by everything on offer. In Melania Trump's world, pigs don't belong in blankets.

I assumed that she was upstairs watching the returns in private or maybe with her husband. But when she didn't text me back, I started to get worried. I left the party and headed up to the private residence.

As I turned into the main hallway, I ran headlong into a tangle of chaos. There was President Trump, surrounded by his gaggle of advisers. The kids were there: Jared and Ivanka, Don Jr., and Eric. So was Bill Stepien. And so were the White House staffers: Mark Meadows, Pat Cipollone, Stephen Miller, Hope Hicks, Kellyanne Conway, and Dan Scavino.

All of them had an opinion, and all of them were speaking at the same time. In the middle was an agitated President Trump, holding some papers and asking questions of nobody in particular as everyone chimed in with answers. Half the group was telling him to go out and lambaste Fox News for calling Arizona earlier in the evening, attack the media for working on behalf of Biden, and talk about voter fraud and the election being stolen. The other half urged a more restrained, presidential posture, arguing that he should say that there were still a lot of ballots to be counted and it was premature to call the race for any candidate. It was chaos and I couldn't tell you who was on what side.

Having seen that show so many times before, I sensed trouble. The yelling and contradictory advice invited confusion, as usual. I thought (once again) how happy I was to no longer be working in the West Wing. But I also felt that it was a time when the first lady's advice would be most needed in the face of all the rancor.

My girl, however, was still nowhere in sight. And there was another problem: the circle of hell orbiting a very agitated president was right between me and the first lady's bedroom door, about fifteen feet away. To get to her, I would have to go past them.

Carefully, I skirted around the edge of the gathering tornado, avoiding eye contact with everyone in the group. When I reached the bedroom door, I knocked a few times, quietly at first but louder at each attempt. I tried to keep an eye on what the group was doing. Some of them had seen me and, if they'd stopped jockeying for half a second, might have figured out what I was up to; no doubt some of them would have tried to prevent me from bringing Mrs. Trump into the fray. Luckily, most of the people in the group were too preoccupied with themselves and airing their own opinions to notice.

Finally, I opened the bedroom door and let myself in, only to find that Mrs. Trump, on the second most important night of her husband's political life and as the White House was enveloped in scenes of shouting jackals and utter chaos, was sound asleep. Maybe she had had enough. I suppose she was just plain tired—literally and figuratively. I marveled that she had managed to tune out all the shouting and commotion just outside her door. I knew by now how much sleep meant to her, but still, I couldn't imagine being asleep at a time like that. Maybe she thought that someone would wake her up if Trump won.

I tiptoed over to her in my heels, which was not an easy feat, considering how thick the carpet was and how dark the room was. I woke her up as gently as I could and explained that the president would likely be giving remarks soon.

Slightly dazed as she awoke, she proceeded to get ready while I sat on the bench at the end of her bed to watch TV. Her son walked in soon after and joined me as I remained glued to the set and yelled updates to her. I've been careful not to discuss him in these pages, but I was so impressed by how calm he was at that moment, even though the election results would determine his future for the next four years, including where he would live. "We'll see what happens" was his mantra that night, which frankly was a better mindset than most of his father's advisers and the other "adults" in the residence had.

Meanwhile, Mrs. Trump kept running between her dressing room and her bedroom, getting ready while peppering me with questions about what states had been called, what states were in play, what her husband was going to say, when she needed to be ready, and what I thought. The whole scene was surreal and frantic and all being done in the dark. For some reason, all of us were so focused on the events at hand that nobody bothered to turn the lights on.

It was no time to sugarcoat things, and in fact, I found myself being more aggressive than I'd ever been with her. After she got ready, we walked to the main area of the residence, where the crowd was still yammering away. We stayed by her door and just watched for a while, it was just such a scene. I finally urged her to jump into the melee and tell her husband to walk onstage, be even keeled, and say the race wasn't over because it was still too early to call. I rarely did that, but I kept repeating myself to her, getting louder and louder each time.

"Please go tell him to stay calm," I urged. "Please tell him there will be plenty of time to fight in the coming days. Biden has really been driving the unifying message home; our president needs to do the same."

I kept stressing to her how important it was that her husband remain calm and presidential, that now was not the time to rant about things or blame others. I told her I thought that a divisive speech, which was being proposed by some of the crew, was the last thing the country or our administration needed. I actually (gently) physically pushed her toward the group at one point. Her husband needed to hear what she had to say. Yet she never said a word. That should have been a preview of what would happen on January 6.

We finally all made our way down to the Green Room, which also served as "offstage announce," meaning that as soon as the president and first lady were announced, they could take the stage. The same group from the residence had gathered, led, of course, by Jared. We were joined by Judge Jeanine Pirro, Rudy Giuliani, and Laura Ingraham, who had been enjoying the party on the State Floor. Why they were there was unclear. What we were going to do was unclear. I was pissed. The president of the United States deserved better than a free-for-all at such a pivotal moment.

I again looked to Mrs. Trump to speak up, to give her husband calming advice, as I had seen her do so many times before. But for some reason she stayed silent. I was confused and irritated by her because I knew she was aware how high the stakes were. Her silence puzzled me until I finally figured it out—or like to think I did. After four years of battles, she, too, was tired of the fight. It had to be why she was asleep when the rest were scrambling and desperate for the president's ear. As was her way, she knew that there was no controlling the ultimate outcome, and she would let the voters decide her fate. While everyone else was gearing up for battle, she was at peace. Her husband and his team were going to do whatever they wanted to in the end—"They will do whatever they want anyway, why fight it?"

They finally took the stage around 2:30 a.m., and it wasn't the speech it could have or should have been. Thirty seconds in, the president blamed "a very sad group of people" for "trying to disenfranchise" his voters. That set the tone. Though Melania got a resounding cheer when her husband thanked her, those who knew her could tell that her fingerprints were nowhere near those remarks.

The press did, of course, pay close attention to her wardrobe—a black power suit—and as usual analyzed it for great meaning and significance. Of course, she hadn't chosen black to make a point but because it was the middle of the night, she was fishing for the closest thing in her closet, and the lights were off. As most women know, when in doubt, grab a black suit or dress, and you are golden. It was just one more example of people trying to decipher the meaning of a woman they didn't remotely understand. I do remember that a couple of days later, she was talking about Biden's acceptance speech and how she was being criticized for not standing next to Trump as Jill had stood next to Biden on election night. She said, "I don't stand next to him because I don't need to hold him up like she does. Can you imagine?" That made me laugh.

AFTER THEY LEFT THE stage, some of my East Wing colleagues and I went back to our offices. It was 3:00 a.m. at that point, and I had been up for almost twenty-four hours. My feet hurt, and we were all exhausted. Some of us lay on the floor in my deputy's office and joked that we should just sleep there and wake up to find out who our president was. One of my junior staffers, a person I had always advocated for and trusted, took a photo of me lying on the floor that night without my knowledge. It was then sent to people in the West Wing advance office, and it made the rounds of several people, including my ex. A week later, I started getting inquiries—and warnings—from reporters that people were saying that an internal investigation was being done because I had been "visibly drunk" on election night and had passed out in a public area. The stories became more outlandish with every reporter I talked to. One version had me passed out in the Blue Room, another in the East Room. One version had me passed out in my office, and one even had me passed out in Mrs. Trump's bed! Fortunately, I managed to speak with every reporter who approached me, and none of them wrote a story because once I explained the facts, it was obvious that the information was another smear job from the Trump White House. The allegations that were lobbed against me snowballed in their severity and included that I had a drinking problem, mental health issues, and problems with addiction, that I was going to lose my security clearance, and that I was suicidal over the breakup.

But as I said earlier, Miranda Devine of the *New York Post* took the bait and wrote one of the cruelest and most dishonest hit pieces I have ever read about anyone, let alone myself. I had my thoughts on exactly who was behind all of the stories being peddled because reporters with a conscience let me know privately what was going on. I will never break that trust, but suffice it to say that a few people were involved, and you wouldn't be surprised as to who. And remember the wellness check that had taken place months earlier? The one that I had been assured would never be talked about because it was a misunderstanding? That, too, was shopped around to the press. I remember Mrs. Trump, referring to the article written by Devine, saying "I hope she wakes up one morning and realizes how awful that was. It was just so mean and so obviously personal. I don't understand it."

What I also found unbelievably cruel was the way that suicide, mental health, drinking, and addiction problems were being lobbed to reporters so casually and in such a way as to humiliate, not help, me. What if I had actually been struggling with addiction or was suicidal? For an administration that had worked so hard on issues of addiction and mental health, it was an irony I still feel disgusted by.

I lined up people who had spent time with me on election night and would attest that I hadn't been drunk. They included a White House physician, an attorney in the counsel's office, Emma Doyle, Rickie Niceta, my personal doctor, an Air Force One flight attendant, another East Wing senior adviser, and Mrs. Trump herself, since I had been with her most of the night. Nothing ever came of the investigation, and I remain grateful to the reporters who were sent a copy of the picture but never used it or wrote any stories. If I were a betting girl, though, I'd lay money that the photo will magically surface after this book is released. I did call out the staffer who took the photo, too. I could not understand how that person could sleep at night, knowing they had done something like that to someone who had been a mentor and a friend. The person never responded, and we've never spoken again.

AS I SAID, THOSE last few months took a real toll on me personally, even physically. I was so angry with so many people in the administration. Though I wasn't surprised that Meadows was going out of his way to rid himself of me, I didn't understand why my ex was being so hateful and deceiving. But I was also angry with the Trump family. They knew me and knew what I had given up for almost six years to support them, and I felt they were allowing the smear campaign to go on. I'm not sure what I expected, and it was a good ego check to be sure, because at the end of the day, I was nothing more than a means to an end, and when I lost my value, I was not worth shielding.

From my perspective, Mrs. Trump changed, too. The woman I knew to be pragmatic and reasonable started to seem less so, especially after the November election. I wasn't always sure that the president truly believed the election had been stolen, as he repeatedly claimed. At least not at first. Before the ballots were even cast, he had telegraphed that he was going to claim voter fraud if he lost. So I wasn't shocked, even if I was disappointed, that he started doing the voter fraud thing that first night. What shocked me was that Mrs. Trump did it in private, too. Whenever we talked about the election, she would say things such as "Something bad happened" or "It looks bad," meaning that the election results weren't legitimate. She was not as into conspiracies as her husband was, so I couldn't believe she was saying that.

Of course, I didn't know who they were listening to in those days. Mark Meadows had gradually tightened the circle of people allowed around the Trumps to just him, his merry band of advisers, and whoever else managed to sneak into the residence: Rudy, Sidney Powell, Matt Gaetz, whoever. Maybe those were the only voices the Trumps were hearing. Part of me cynically believed that Mrs. Trump was going along with the voter fraud claims only because she knew she was going to be leaving the White House and would have to live with the guy afterward. Maybe she didn't want to get on his bad side over it, since there would be far fewer people around to insulate her from him. Whenever the subject came up, I gingerly suggested that there were some cases of voter fraud in every election, but I wasn't so sure that there had been some grand conspiracy to defraud Trump. She didn't want to hear that. "It's bad," she said over and over, as if she were privy to some info I didn't have. I threw up my hands.

BUT AT SOME POINT in December, I think Mrs. Trump saw where it was all heading and started packing up things. And as I touched on previously, she put her photographer on overtime duty to assemble a bunch of albums. I was horrified by the hours the photographer was billing, since I had to approve them, and the Office of Administration kept questioning me about them. To make it worse, I never did see the albums. But the photos were, I suppose, a source of comfort to the first lady.

Ultimately, I think Mrs. Trump had a mixed view of the election result. Perhaps she didn't like the bruise to her ego over her husband having lost, but maybe she was also glad to go back to a life of privacy. I was also surprised that she went along with Trump's plan to snub the Bidens. I tried suggesting a couple of times that she welcome Jill Biden to the White House or have her over for tea, as other first ladies had done. She would always say, "Let me think about it" or "Let's see what the West Wing will do." Which meant no. And when exactly did she decide to start following the West Wing's lead?

THEN JANUARY 6 HAPPENED. As I said in the beginning of this book, I had an uneasy feeling about the day. I was at home in DC, watching everything unfold on TV. As you have likely gleaned, I was almost completely walled off from my colleagues in both the East and West wings at that point. It was part exhaustion, part not giving a shit anymore, and a whole lot of self-preservation. I was operating under a veil of total paranoia and was disheartened by and disappointed in both the president and first lady. I was also grappling with the fact that my ego had gotten out of control enough for me to think that I was special in the Trump circle, and shocked by how far I had "fallen." Remember, there was a time that the president had told me "everyone just loves you," so I couldn't stop wondering what the hell happened. January 6 was a truly devastating day for the country, but looking back now, it was bound to happen. All of it. The insurrection and my resignation. January 6 was a day of reckoning for me, though as I watch things now it seems that I am in the minority on that way of thinking.

After I resigned, my anxiety didn't melt away like I thought it would. While I had no regrets about my decision, it was still hard to be cut off so quickly by people I had been in the trenches with for years. I am a human being, and it hurt. The days that followed were hard, like what I imagine leaving a cult would be like. I tried to avoid the news. I walked Ben a lot and I instructed my close circle of friends not to tell me anything about Trump World in the hopes that it would help me move on, and it did to some degree. Still, my anxiety and feelings of anger took a long time for me to reckon with. There is a period of guilt, of sadness, of deprogramming, and then you start to rebuild. To be honest, my feelings about my experience are still a big ball of a mess that I am not sure I will ever fully untangle.

Epilogue

Who looks outside, dreams. Who looks inside, awakes.

—CARL JUNG

As I wrote earlier, I haven't spoken to many of my old colleagues since I left. And I have not spoken to Mrs. Trump since the day I resigned, which took some getting used to as we had talked or texted almost every day for years.

After they left the White House, Mrs. Trump sent all of her employees personalized letters thanking them for their service, along with a candid photo with them that she had signed. I know that because the gift was my idea and I wrote the letters, which were filled with personalized details as well as their job titles. Not that it matters at this point, but I received no gift, and my letter was so vague and cold that a doorman would have been offended. It read, "Dear Stephanie, Thank you for your service to the American people as a member of the office of the First Lady. I hope you look back on your time at the White House as cherished, knowing you helped serve our country. I send my best wishes to you on your next endeavor."

Quite the "Goodbye, and good luck," right? It's the equivalent of a "Dear John" letter for the workplace. I'm not sure that was done because I resigned on January 6 or because my resignation went public in the press and she felt betrayed. But I know her well enough to know that she knew what she was doing with that letter.

*   *   *

IT HAS OCCURRED TO me as I've been writing that I seem to be blaming everyone but myself for how things turned out for me in the White House, especially in the last six months. According to me I was the victim of covid, of Meadows and his people, of my ex, of the former East Wing chief of staff, of some of my own East Wing staff, of some West Wing senior staff, of the president, and even of the first lady at the very end. And although the stories I have laid out are all true and it was very much a perfect storm of certain personalities coming together in opposition to me, I don't feel that I am a victim who did no wrong. It is my fervent belief that when you are the common denominator in situations like this, you need to look within and determine where your own responsibility lies. People need to hold themselves accountable to situations so that they can learn from them and apply them in the next chapter of life, and that includes me.

I think the first part is obvious: I became heady with power. I got cocky. You get inside the walls of the White House, the most important building in the country and arguably the world, and you are catered to like nowhere else. You go in wanting to help the people of the United States, but I don't think many people in the Trump administration left there as the best versions of themselves; I know I did not.

Second, when you work in a *Hunger Games*–style environment like that, another instinct takes over: instead of focusing on getting productive work done, you just want to survive. So you do whatever you have to do to make that happen, including making compromises with yourself and your morality that don't sit well with you. I was guilty of that, too.

At the same time, though, I did think somebody needed to stick around to look out for Mrs. Trump. I was loyal to her personally, and I didn't want her to be staffed by incompetent or untrust-worthy people who didn't have her best interests at heart. And as she had most always been good to me, I felt gratitude. But her apathy in response to the January 6 riots made it hard for me to stay at the very end.

I also turned a blind eye toward my own falling into a trap I saw over and over again: believing I was a trusted and valued member of Trump World. The plain truth is that most of the Trump fam-ily dismisses and cuts people from their lives on a whim. They demand total loyalty, but they are loyal to no one. I don't blame them, to be honest. They are businesspeople, and business should not be personal. Some people learned that once and walked away; others kept going back for more, and there are many who are still doing it. I allowed my ego to grow in such a way that I never con-sidered that the Trumps would allow me to be treated poorly. I put myself onto the same level as Hope Hicks, Dan Scavino, even Javanka, and that was ludicrous. Mrs. Trump did defend me when she could, and privately she always told me of her anger on my be-half, but I'm not sure it ever went farther than that, and I wrongly expected that it should have.

Finally, and most importantly, I should have spoken up more.

I think a lot about my grandfather, my father's father. As a young girl, I spent summers with my dad's family at their farm in a small town called Fruita on the western slope of Colorado. As I grew older, I found great joy in debating my grandfather, a staunch Republican, on a myriad of social issues and telling him I planned to be a Democrat because clearly I didn't agree with so many of his positions. He would always just smile and tell me it was okay to be affiliated with any party I wanted, as long as I par-ticipated in the political well-being of and service to our country.

Looking back, no matter the ire it would have caused or the reactions within the White House, I should have had the guts to speak up about things I thought were wrong, even if it would have made my time in the administration shorter. I was weak not to do that, and because of my fear, I didn't serve the country well, which is my biggest regret of all.

The scarier thing to think about, though, is the well-being and future of our country. I've been a Republican most of my life, and I've worked all my career to support constitutional and conservative values. But today, being a constitutional conservative doesn't seem to be enough to be a "good Republican"; what seems to matter today is blind loyalty to an ex-president who still won't admit he lost. No matter the party, we should be loyal to this country, not to any one man or woman. It saddens me to see so many Republicans allowing one man to have the perceived power to rule our party; that is called an autocracy and that is not how this country was founded. As I write these words, I am sitting far removed from DC and have watched my party vilify a woman for simply speaking her opinion. To relieve Congresswoman Liz Cheney of her leadership role because she "continues to live in the past" while the forty-fifth president sends daily missives out about how the election was stolen is just batshit crazy to me. The Republican men in leadership claim to want to stay "on message" and look to the future, all the while they do whatever they can to stay on the good side of a man who is focused only on the past, on revenge, and on himself. President Trump gave a voice for many people who felt forgotten or unheard. But what started out as a positive change in the status quo has descended into a frenzy of anger and violence. I believe that the Trump administration put into place many excellent policies that I hope will continue, but those are Republican policies, not Trump policies. He does not own them, and I firmly believe that we as a party can move them forward without the divisive, scandal-laden drama of the years we were in office. Simply put: the Republican Party is not one man.

*   *   *

AS I WROTE IN the beginning, this book will probably piss a lot of people off, and I have had a year to prepare for it and almost six years to grow a very thick skin. Some on the right will say I'm a traitor, and some on the left will say it's too little, too late because I willingly worked for a "monster." People I once admired or called friends will attack me, and I laid out everything I imagine Trump World will say. I say it's just a book, with stories about a time in our history that will be talked about for years to come. I hope you laughed a bit, and I hope you at least understand a little more.

I read a quote the other day that I think sums things up nicely:

I don't think you get through life without any regrets, but you can create some purpose from it.

Acknowledgments

I want to thank my team at HarperCollins, led by my editor, Jonathan Jao. It is not lost on me that you took a chance here. The process was difficult for me emotionally and the support and confidence I received from you, David, Tina, and Theresa is what allowed me to shape this book into what I wanted it to be. Thank you for always listening and keeping me honest with myself.

I cannot express enough gratitude to Matt Latimer, founding partner at Javelin and my agent. You believed in what I wanted to do and kept me calm and grounded while I was doing it. Also thank you to Dylan Colligan, who arguably had the hardest job of all when it came to research and keeping it all organized.

Thank you again to the family and close friends who were aware of what I was doing, and remained patient with my schedule, my neurosis, and my insecurities as I went through this journey. I would name each and every one of you if I didn't think there could be some kind of retaliation. I love you all.

About the Author

Stephanie Grisham started at the White House on January 20, 2017. She served as White House press secretary and communications director from 2019 to 2020. She also worked as communications director and chief of staff to First Lady Melania Trump. Born in Colorado, Grisham lives in Kansas and Washington, DC, and is the mother of two boys, Kurtis and Jake.